Wastewater Analysis for Substance Abuse Monitoring and Policy Development

Wastewater Analysis for Substance Abuse Monitoring and Policy Development

Jeremy Prichard
Wayne Hall
Paul Kirkbride
Jake O'Brien

CRC Press is an imprint of the
Taylor & Francis Group, an informa business

First edition published 2021
by CRC Press
6000 Broken Sound Parkway NW, Suite 300, Boca Raton, FL 33487-2742

and by CRC Press
2 Park Square, Milton Park, Abingdon, Oxon OX14 4RN

© 2021 Taylor & Francis Group, LLC
CRC Press is an imprint of Taylor & Francis Group, LLC

Reasonable efforts have been made to publish reliable data and information, but the author and publisher cannot assume responsibility for the validity of all materials or the consequences of their use. The authors and publishers have attempted to trace the copyright holders of all material reproduced in this publication and apologize to copyright holders if permission to publish in this form has not been obtained. If any copyright material has not been acknowledged please write and let us know so we may rectify in any future reprint.

Except as permitted under U.S. Copyright Law, no part of this book may be reprinted, reproduced, transmitted, or utilized in any form by any electronic, mechanical, or other means, now known or hereafter invented, including photocopying, microfilming, and recording, or in any information storage or retrieval system, without written permission from the publishers.

For permission to photocopy or use material electronically from this work, access www.copyright.com or contact the Copyright Clearance Center, Inc. (CCC), 222 Rosewood Drive, Danvers, MA 01923, 978-750-8400. For works that are not available on CCC please contact mpkbookspermissions@tandf.co.uk

Trademark notice: Product or corporate names may be trademarks or registered trademarks and are used only for identification and explanation without intent to infringe.

Library of Congress Cataloging-in-Publication Data
Names: Prichard, Jeremy, author.
Title: Wastewater analysis for substance abuse monitoring and policy development / Jeremy Prichard, Wayne Hall, Paul Kirkbride, Jake O'Brien.
Description: First edition. | Boca Raton : Taylor and Francis, 2021. | Includes bibliographical references and index.
Identifiers: LCCN 2020025683 (print) | LCCN 2020025684 (ebook) | ISBN 9780367132903 (hardback) | ISBN 9780429025709 (ebook)
Subjects: LCSH: Substance abuse. | Sewage–Sampling. | Water–Sampling.
Classification: LCC RA567 .P75 2021 (print) | LCC RA567 (ebook) | DDC 363.72/8–dc23 LC record available at https://lccn.loc.gov/2020025683
LC ebook record available at https://lccn.loc.gov/2020025684

ISBN: 978-0-367-13290-3 (hbk)
ISBN: 978-0-429-02570-9 (ebk)

Typeset in Palatino
by Newgen Publishing UK

Contents

Preface

The analytical chemistry that drives wastewater analysis (WWA) can be applied to study a wide variety of topics, including human impacts on the environment and numerous dimensions of human health. At the time of writing, research had just commenced to test the efficacy of WWA to study the COVID-19 virus.

Our interest in this book is in exploring the contexts where WWA may be able to assist in reducing the social burdens caused by illicit drugs and other substances. Like other scholars, we are convinced that multi-disciplinary research is essential to address complex 21st century problems. Consequently, in each chapter of this book we have attempted to encourage policy makers and academics from multiple disciplines to consider how they may be able to use WWA with other methods to develop new and useful collaborations.

Acknowledgements

We thank Barbara Knott and Danielle Zarfati, CRC Press/Taylor & Francis Group, for their kind and practical support in completing this project. We are also grateful to Dr Ben Tscharke, Queensland Alliance for Environmental Health Sciences, and Olumayowa Adesanya and Dr Michael Guerzoni, University of Tasmania, for their comments and suggestions on drafts of our chapters.

About the Authors

Jeremy Prichard is an Associate Professor of Criminal Law at the University of Tasmania and an Adjunct Associate Professor at the University of Queensland. His earlier professional roles included appointments at the Australian Institute of Criminology, the Queensland Department of the Premier and Cabinet, and the Queensland Department of Communities, Aboriginal and Torres Strait Islander Partnerships.

Wayne Hall is a Professorial Fellow in the National Centre for Youth Substance Use Research and the Queensland Alliance for Environmental Health Sciences, both at the University of Queensland. He has Visiting Professorial appointments at the London School of Hygiene and Tropical Medicine and the National Drug and Alcohol Research Centre, University of New South Wales (UNSW).

Paul Kirkbride is Strategic Professor of Forensic Science at Flinders University in South Australia. Prior to that academic appointment, he was for many years an operational forensic scientist and senior manager at Forensic Science SA, Manager of Business Programs at the National Institute of Forensic Science, and Chief Scientist with the Forensic and Data Centres portfolio of the Australian Federal Police.

Jake O'Brien is a Research Fellow at the Queensland Alliance for Environmental Health Sciences (QAEHS) at The University of Queensland (UQ). He has a keen interest in wastewater-based epidemiology and his PhD focussed on refining the uncertainties and expansion of wastewater-based epidemiology for assessing population exposure to chemicals (conferred in 2017, UQ).

chapter one

Measuring Strategies to Counter the Harms of Substance Use

A Global Overview

INTRODUCTION

> If progress is to be made on the prevention and treatment of drug problems, new quantitative and qualitative research methods need to be applied in large population areas, methodologies that do not conform readily to traditional experimental and evaluation designs.
>
> **(Babor et al. 2010b, 97)**

This book is about how to measure substance use and thereby evaluate policies intended to reduce drug-related harm. The book can be précised in four main points.

First, the misuse of psychoactive substances, including illicit drugs, alcohol and tobacco, constitutes a very significant and complex health and social burden in all parts of the globe.

Second, at international, national and local levels, there are combinations of strategies used with the aim of reducing this burden. These aim to reduce the supply and the demand for psychoactive substances. Harm reduction strategies attempt to reduce the harms associated with the use of psychoactive substances without necessarily reducing demand or supply.

Third, it is often difficult to assess to what extent these strategies achieve their desired effects, or produce unintended consequences. Assessments of policy efficacy draw upon a wide variety of information, including intelligence and quantitative and qualitative data on drug use.

Fourth, a new method has emerged to measure the whole population consumption of psychoactive substances. Called *wastewater analysis*, the method relies upon separation science to detect traces of specific substances in sewage water and uses these traces to estimate per capita consumption of the substance.

This book is responding to the call made by Babor and colleagues (2010a) quoted above. We ask whether wastewater analysis can be incorporated into research designs to meaningfully assess the impact of policies that aim to reduce the harms of psychoactive substance use. This is the book's central question. We answer it by examining the capabilities of wastewater analysis and critically discussing whether the method may be put to good use in evaluating policies at international, national and local levels.

The book's focus is weighted towards illicit drug use because of the special impediments that researchers face in measuring the use of drugs provided by an underground (black) market. However, we also discuss the potential use of wastewater analysis in assessing alcohol and tobacco use from criminological and epidemiological perspectives.

The authors' backgrounds are in chemistry, epidemiology, criminology, law and policy analysis. We have attempted to avoid biases towards particular fields as best we can to produce a book that is 'transdisciplinary' in the sense Henry (2012) used this term, collaborative research relationships between biological and social scientists and experts in water management, health and law enforcement. Commentators from completely different viewpoints are increasingly recommending diverse collaborations to address the world's most complex problems, including substance misuse. These sentiments have been voiced by scientists (Ledford 2015), epidemiologists (Babor et al. 2010b), criminologists (Henry 2012) and wastewater researchers (e.g. Castiglioni, Vandam and Griffiths 2016; Prichard et al. 2017).

Terminology

Consistent with the social science literature on the topic, we have adopted the term 'psychoactive substances' to collectively refer to illicit drugs, alcohol and tobacco. We refer to their use as 'substance use'. In some sections we use the term 'drugs' interchangeably with illicit drugs. However, 'drug' should be taken to include alcohol and tobacco. In the case of the non-therapeutic use of pharmaceuticals, our key term is 'pharmaceutical misuse'.

'New psychoactive substances' is a phrase that refers to a rapidly expanding list of synthetic compounds that mimic the effects of widely used illicit drugs but which fall outside of current definitions in criminal law. The book does not discuss this class of substances extensively for reasons provided in Chapter 2 (2.2.4). 'Psychoactive inhalants', such as glue, petrol, paints and so on are also not considered.

Normally, the literature refers to alcohol 'use', the 'use' of illicit drugs and so forth. We employ the same terminology except where wastewater data are concerned. Following a convention that acknowledges the inability of wastewater data to directly infer anything about the number of users and their patterns or frequency of use, we refer to wastewater data in terms of what they tell us about the total 'consumption' of substances in a given population.

'Wastewater analysis' has at times been called 'sewage-based epidemiology' and more recently 'wastewater-based epidemiology' (WBE) (e.g. Castiglioni and Vandam 2016). As implied above, we prefer wastewater analysis (WWA). We think this is a simpler term. We also think that terms like WBE denote *how* wastewater data might be used – to epidemiological ends – rather than what wastewater analysis *is* (mainly chemical analyses of drug residues and metabolites in wastewater samples). Additionally, since language expresses meaning, we are concerned that terms like WBE may discourage non-epidemiologists from engaging with this new field, thereby undermining efforts to incorporate expertise from broader social science and other research domains.

Structure of This Book

Chapter 1 sets the framework for the book by introducing three interlocking themes: the problems raised by substance misuse; the strategies available to address these problems; and the methods available to assess whether these strategies are working.

Cutting across those themes are three tiers – macro (international), meso (national) and micro (local). Section 1.1 accordingly explains the costs and harms of substance at the individual, national and global level. The main stratagems to counter the harms associated with substance use are overviewed in Section 1.2, including international, national and local efforts. The focus of Section 1.3 is on assessing the efficacy of these different tiers of strategies.

Chapter 2 explains WWA from a social scientist's perspective. The aim is to equip the reader with a clear sense of the strengths and weaknesses of WWA compared to other sources of information on substance use. It answers common questions about WWA, such as how it copes with the dumping of illicit drugs down sewers. The chapter gives the reader a sense of the WWA field that is transparent about the capabilities of WWA, the ethical issues it may raise and potential interactions with experts from other areas. Because the chapter describes the key stages in WWA studies, it is also a useful resource for readers who may be considering participating in or managing a WWA project.

Chapters 3, 4 and 5 are *investigative.* They deal with the macro, meso and micro tiers referred to above. The chapters have the same format. They start with the limitations of the current sources of information and methods for assessing the efficacy of strategies. Then they dispassionately examine whether WWA might compensate for some of those limitations. These chapters consider a spectrum of research designs that range from: the simple use of WWA data to estimate population drug consumption, time series analyses of trends in drug consumption, and incorporating WWA data into evaluations of natural policy experiments, intervention studies and quasi-experimental designs (see Babor et al. 2010b, 8, 99).

Global issues are at the centre of Chapter 3, which surveys key international agreements and agencies. The principal topics examined are the value of WWA in developing countries and the potential for WWA to provide comparable global metrics on consumption of alcohol and tobacco.

Chapter 4 deals with national applications of WWA. It devotes particular attention to the estimation of the extent of rural substance use, especially in large countries like Canada, the USA and Australia. It uses recent case-studies to consider the broader implications of WWA for small communities, which often lie in a 'data shadow' because of the difficulties in using traditional social science survey research methods to estimate their drug use. Paradoxically, there may be empirical *advantages* to conducting WWA in towns and small communities that are isolated. The ethical issues raised by this type of research are also discussed.

In Chapter 5 the utility of WWA in micro contexts is deliberated. We discuss technical challenges relating to sampling from small sewers with highly variable water flow. The chapter devotes particular attention to the problem of substance use by prisoners and corrective services staff. Using WWA prison studies previously undertaken (e.g. Brewer et al. 2016), the chapter explores how WWA intervention studies could evaluate the effectiveness of supply or demand reduction strategies. Schools and workplaces are the other micro settings that raise the highest level of ethical complexity for WWA research.

The concluding sixth chapter is *speculative* because it considers what future scientific applications of WWA might mean for the work of criminologists and epidemiologists and for evaluations of drug policy. Future risks for WWA are examined, including technical changes in sewage treatment and the mismanagement of ethically sensitive WWA studies.

1.1 The Problem of Substance Misuse

We start with an overview of the problem of substance use to explain why the measurement of drug use is so important. We also explain why measurement is often difficult. The full implications of substance misuse for the global community are impossible to gauge. Even at the individual level, causation can be complex and ripple effects hard to disentangle from the effects of other factors in the life-course. For instance, how can the long-term psychological implications of parental drug dependence on children be delineated? Or the premature death of a spouse from a tobacco-related disease?

As we widen our scope from the individual to the community, the nation, the region and the globe, the complexity of harm increases by orders of magnitude. Substance use causes significant harm to public health and wellbeing and places a large burden upon criminal justice agencies (WHO 2017b; UNODC 2016b). Illicit drugs produce different problems to alcohol and tobacco because illicit drug markets are a major source of income for global organised crime groups.[1] As such, they have a distinct geo-political relevance.

> Organized crime feeds on instability and the weak rule of law, shows violence, and grows strong with help from corruption and money-laundering. It threatens the safety and security of communities, violates human rights and undermines economic, social, cultural, political and civil development of societies around the world.
> *(UNODC 2016b, 32)*

1.1.1 *Harms for the Individual and Community*

Drug use can adversely affect people's physical health, personal safety, mental health and social wellbeing (Loxley et al. 2004). Three variables that affect harms are dose, patterns of substance use and the route of administration (see Babor et al. 2010b, 19). Regarding dose, psychoactive substances differ in terms of their toxicity and their potential for harm. Substance use behaviours encompass a cluster of related but discrete issues: regularity of use (including substance dependence); quantity consumed at any one time (even if only occasional); and poly-substance use. The main routes by which individuals administer substances to themselves (or another) are orally, by intravenous injection or by smoking.

These variables may affect harms experienced in a wide variety of ways. For example, a person who rarely uses illicit drugs may experience acute harm because they injected a large dose of a toxic drug (possibly in combination with other substances). Chronic health or behaviour problems may be caused by persistent use over a long period of a drug with a low level of toxicity.

Toxic effects include physical health consequences such as drug overdoses, chronic illnesses and effects on fetuses in utero. *Intoxication* includes harmful consequences that are temporally linked to the period when a user is intoxicated, e.g. an injury to the user or another person (e.g. drowning, criminal violence or causing traffic accidents). Harms associated with substance use *dependence* include chronic illness, involvement in the criminal justice system and consequences labelled as 'role failures' in more important social roles, such as being a parent, spouse, student, employee and so forth.

How can psychoactive substances be ranked according to their potential for harm? Scholars have taken different approaches to this difficult question, but studies have consistently found that alcohol is among the most harmful psychoactive substances (e.g. Nutt et al. 2007; see WHO 2014). Unrecorded alcohol use, including those that are sold for profit but are unregulated (or illegal), may pose greater health risks than regulated supplies because of unsafe production and contaminants (Rehm et al. 2014). Not surprisingly, tobacco ranks very highly indeed on health-related harm (e.g. Hall, Room and Bondy 1999). Risks associated with the misuse of pharmaceuticals, such as benzodiazepines and buprenorphine, have also been underscored (Nutt et al. 2007). Among illicit drugs, a substantial portion of the health harm has been linked to opioids because of the high risk of fatal overdose (Degenhardt et al. 2014). While less is known about the harms arising from the use of new psychoactive substances, they have been linked to deaths and other adverse health outcomes (Schifano et al. 2016).

1.1.2 Harm at the National and Global Scale

How large are the markets for tobacco, alcohol and illicit drugs? Chapter 3 examines estimates of the prevalence of the consumption of psychoactive substances worldwide and links this critical information with the methods used to monitor substance use. We note that the market for illicit psychoactive substances is very large. The UNODC (2016a) estimated that in 2014 1 in 20 people aged between 15 and 64 years used an illicit drug at least once; this translates to approximately a quarter of a billion people. Best estimates on alcohol consumption suggest that in 2016 persons 15 years and older drank 6.4 litres of pure alcohol per capita (WHO 2017a). The WHO (2017b, 59) has estimated that there are approximately 1.1 billion smokers globally.

1.1.2.1 Economic Perspectives

In a number of countries estimates have been made of the direct and indirect economic cost of illicit drug use by taking into account the economic value of lost productivity, premature deaths and public spending on law enforcement, prisons and health. An analysis by the Office of National Drug Control Policy (ONDCP 2004) suggested that illicit drug use cost the USA US$180.9 billion in 2002. The same pattern emerged each year between 1992 and 2002 with losses in productivity the largest economic cost. The second category used by the ONDCP (ONDCP 2004, xi) combined spending by the criminal justice system, effects of crime on victims and 'a modest level of expenses for administration of the social welfare system social welfare spending'. This category accounted for approximately 20% of costs. The third category, health-care costs, accounted for the remaining 8.7%.

Other studies have adopted broader parameters by calculating the cost of alcohol and tobacco in combination with illicit drugs. Rehm et al. (2006) estimated that the cost of psychoactive substances for Canada in 2002 was CA$39.8 billion. This study attributed 61% of economic costs to losses in productivity, 22% to health care, 14% to law enforcement and 3% to other factors. They estimated that 20.7% of the CA$39.8 billion (CA$8.2 billion) was attributed to illicit drug use. The remainder was attributed to tobacco (42.7%, CA$17 billion) and alcohol (36.6%, CA$14.6 billion).

Similar estimates were produced in Australia (Collins and Lapsley 2008). Psychoactive substances were calculated to cost the country A$55.2 billion in the 2004–05 financial year. Just under 15% of the total was attributed to illicit drug use (A$8.2 billion) and a further 2% to a combination of illicit drugs and alcohol (A$1.1 billion). The bulk of the costs were attributed to tobacco (56.2%, A$31.0 billion) and alcohol (27.3%, A$15.1 billion), although some of these costs were offset by government income from taxation (see similarly, USNCI and WHO 2016).

These approaches examine the economic impacts of substance abuse but the economic costs of other harms may be extremely difficult to measure. For example, it is feasible that in some countries the illicit drug market distorts prices of legitimate goods, such as real estate, negatively influences foreign investment and exacerbates unequal wealth distributions (UNODC 2017).

1.1.2.2 Epidemiological Perspectives

The seminal estimates of the worldwide health burdens of hundreds of illnesses and injuries is the Global Burden of Disease (GBD) that is now led by a consortium of universities and the World Health Organization (WHO). The GBD research has undertaken three main data collection phases related to the years 1990, 2005 and 2010 (Degenhardt et al. 2013). Methods underpinning the GBD include assessments of the strength of the evidence on relationships between (a) diseases, injuries and risk factors and (b) years of life lost due to premature mortality and lived with disability.

The GBD's core metric is disability-adjusted life years (DALYs) which measures the loss of healthy life, either through premature mortality (years of life lost) or through disability (years of life lived with disability) (Degenhardt et al. 2013). A single DALY represents a single year of health *lost* to disability. This measure is inherently individual-centric; DALYs do not attempt to comprehensively measure broader harms to families or society, or their economic consequences. The GBD metrics are also subject to the limitations of data availability (or the lack thereof) in some regions. Notwithstanding, the GBD data are useful in prioritising global efforts to alleviate disease burden because they measure the harms associated with so many diseases and injuries across 21 regions of the world.

The GBD projects show that psychoactive substances contribute significantly to global disease. Worldwide, tobacco accounted for 6.3% of all DALYs in 2010 and alcohol accounted for 3.9% (Lim et al. 2012). Analyses on illicit drugs were constrained by deficiencies in the statistical evidence on the *strength* of relationships between drug use and potential harms (Degenhardt et al. 2013). For this reason, only some dimensions of illicit drug-related harm were incorporated into modelling. These included the relationships between: injecting drugs and HIV, hepatitis C and hepatitis B; and suicide and dependence on cannabis, opioids, cocaine or amphetamines.

Illicit drugs were still estimated to account for 20 million DALYs in 2010, which represents 0.8% of all DALYs worldwide (Degenhardt et al. 2013). This ranked illicit drugs as the 19th leading risk factor overall (Degenhardt et al. 2013) and a greater contributor to the global burden of disease than, for example, all maternal factors combined (Murray et al. 2012). Of the 20 million DALYs related to drug use, 13.7 million affected males. Geographical differences were stark. The largest proportion of DALYs attributable to drugs were in Australasia and parts of North America defined as high-income; the lowest occurred in central and western sub-Saharan Africa (Degenhardt et al. 2013). However, confidence in geographical comparisons was limited by the paucity of prevalence data in many poorer countries.

1.2 Overviewing Strategies to Counter the Harms of Psychoactive Substance Use

A common way of classifying strategies that aim to reduce harms related to psychoactive substances is whether their objective is to reduce market *demand* or *supply*. These classifications are used in policy platforms at the national (e.g. CA-DH 2017), regional (e.g. EMCDDA 2015) and international (UNODC 2016b) level. Some documents also include a third category, namely strategies that aim to reduce the *harms* connected to psychoactive substance use, such as the spread of blood-borne viruses through injecting drugs (e.g. CA-DH 2017; UNODC 2016b), without necessarily attempting to prevent use.

Substance use 'policies' can be defined as any law or government programs that attempt to influence citizens' use of substances and to reduce the consequences of that use (Babor et al.m 2010b). This means that policy incorporates government business in multiple portfolios, for example:

(a) medical infrastructure required to treat acute and chronic illnesses related to substance use;

(b) regulatory mechanisms governing pharmaceuticals and the legal age at which tobacco and alcohol can be purchased;
(c) educative programs designed to prevent or delay the onset of drug use (e.g. first use of drugs or tobacco);
(d) tax agendas to reduce heavy consumption (e.g. alcohol) or encourage desistance (e.g. tobacco taxation);
(e) social welfare and health services to address dependence on psychoactive substances;
(f) enforcement of criminal laws proscribing the production, trafficking and use of illicit substances, and, unlicensed commercial activity involving alcohol, tobacco or pharmaceuticals; and
(g) enforcement of regulatory offences governing the use of motor vehicles while affected by psychoactive substances.

Further details about particular strategies will be discussed in later chapters. Babor et al. (2010b, 262–9) summarise 43 types of interventions that target illicit drugs. They classified these into five main categories – providing yet another way to appreciate variations in policy.

First, were strategies that target particular groups of people, such as school children, families or communities. In this category one could include prisons and workplaces.

Second, were strategies or services designed to encourage behaviour change, for instance through psychosocial treatment, peer self-help organisations, needle exchange programs or pharmaceutical interventions (e.g. methadone to reduce heroin use).

Third, were regulatory interventions that manage over-the-counter sales, prescriptions for pharmaceuticals and so forth.

In the fourth category Babor et al. (2010b) placed amendments to criminal laws or policing policies, which include court diversionary practices, drug courts, sentencing options and models of legalisation or decriminalisation of personal drug use.

Their final category was supply control interventions. This is the same as 'supply reduction' mentioned above. Supply reduction includes law enforcement operations to disrupt criminal networks and border control to intercept the importation of drugs or their pre-cursors. Supply reduction is a goal in both domestic and international settings (UNODC 2016b). Examples of multilateral interventions include efforts to eradicate agricultural production of cocaine and to provide farmers with alternative (legitimate) markets and sources of income.

For illicit drugs, the global policy landscape is dominated by three UN treaties: the *Single Convention on Narcotic Drugs* (1961), the *Convention on Psychotropic Substances* (1971) and the *Convention against Illicit Traffic in Narcotic Drugs and Psychotropic Substances* (1988). These sources of international law provide a mandate for the United Nations to regulate the world's drug market. The International Narcotics Control Board (INCB) regulates medical use of controlled substances such as opiates and cannabis via a system of estimates and approvals on their production and trade. The UN Office on Drugs and Crime (UNODC 2016b, 12) encourages member states to pursue a 'balanced and comprehensive approach' with a mix of demand, supply and harm reduction strategies. Harm reduction objectives of the UNODC overlap with those of other UN bodies. These include the WHO and the Joint United Nations Programme on HIV/AIDS (UNODC 2016b).

The situation regarding alcohol and tobacco is different because they are legal commodities and there is heavy corporate involvement in their legal production and supply (although both can also be trafficked illegally, as discussed in Chapter 3). At the international and regional level there are a variety of agencies that pursue agendas relating to alcohol and tobacco.

The key international document for alcohol is the *Global Strategy to Reduce the Harmful Use of Alcohol* (WHO 2010). Examples of alcohol strategies commonly employed around the world include: tax excises, restrictions on alcohol advertising, retail regulations (e.g. restricting hours of sale), drink-driving laws, treatment and early intervention, and education programs (Babor et al. 2010a; WHO 2017). In Europe the WHO facilitates networking between countries to monitor progress in implementing the *European action plan to reduce the harmful use of alcohol 2012–2020* (WHO 2012).

The WHO also plays a central effort in strategies targeting tobacco alongside national and regional bodies. The *WHO Framework Convention on Tobacco Control* (adopted 2003) sets out demand and supply reduction strategies for member states to reduce the burden of disease attributable to tobacco (WHO 2003). Articles 6 to 17 list measures relating to, among other things: demand reduction for dependence cessation; education programs; restrictions on tobacco production, advertising and sales; protecting third parties from tobacco smoke; tax excises; and supply reduction for illicit tobacco products (WHO 2003). In 2016 the WHO (2017b) estimated that 121 countries (two-thirds of all nations) had implemented at least one of the best-practice tobacco policy interventions.

Aspects of the international and regional policy frameworks are returned to in Chapter 3. Chapter 4 focuses on the national policy context. Chapter 5 will discuss policy platforms that exist at the institutional level.

1.3 Current Methods for Evaluating the Efficacy of Strategies

In most books on psychoactive substances, data about the nature and extent of substance use are discussed in relation to the scale of harm arising from their use. This book approaches data differently. Sections 1.1 and 1.2 have briefly overviewed the harms associated with substance use and major strategies contained within policy platforms. This third section explores how data relate to these two topics, namely: how data on substance use are generated, and how they are used to evaluate strategies and provide evidence to inform policy.

This section deals with illicit drugs, alcohol and tobacco simultaneously. The rationale for doing so is that many countries have policy platforms that cover all psychoactive substances (e.g. CA-DH 2017). Additionally, as the section explains, there are strong similarities between the three classes of substances in terms of data generation.

Policy on substance use draws on many types of research. Clinical and medical research, as one example, provides evidence about the efficacy of medical treatments and pharmaceuticals.

Studies that evaluate strategies, sometimes called policy *evaluation*, bear similarities to policy evaluations in other fields of research in terms of the use of statistical methods and quasi-experimental designs. However, evaluation research differs from research in being 'an applied inquiry process for collecting and synthesising evidence' on the effects of a service, program, or policy against the goals they set out to accomplish (Babor et al. 2010b, 100). Evaluation research might have implications for resource allocation, including whether to implement new strategies, phase out ineffective strategies, or direct more resources to one strategy than another. Scholars argue that the difficulties associated with evaluation research are not well appreciated, especially at the international level (Hall 2018).

Evaluations may be undertaken as discrete, one-off studies by research institutes and centres, government-funded agencies or academic institutions. Data from ongoing monitoring systems may be used in evaluations. One of the particular strengths of monitoring approaches is that they provide data collected using a consistent method over long periods of time. The *National Survey on Drug Use and Health*, for example, is the lead source of survey data on substance use in the general population of the United States that has been collecting data since 1971.[2]

A hallmark of evaluation research is that it often employs mixed-methods approaches so that the strengths of some methods compensate for the weaknesses of others (Campbell, 1969). Mixed-methods are needed because the phenomena in question cannot be examined by a single approach. Consider the prevalence of illicit drug use: a clandestine behaviour carrying social stigma. Every acceptable source of data – surveys, official statistics on arrests or drug overdoses, qualitative data and so on – are drawn on to produce the best estimate of the prevalence of drug use. The term 'triangulation' is used in substance use research in reference to the process of comparing estimates produced by different data sources. Given the meaning of this word in surveying, it may need to be used cautiously to avoid over-stating the precision of evaluation research.

Substance use changes over time and varies between places, something explored in various ways in Chapters 3, 4 and 5. We can briefly support this claim by pointing to findings that (a) the global burden attributed to opioid overdoses increased between 1990 and 2010 (accounting for population growth and other factors) (Degenhardt et al. 2014); and (b) there are significant differences in smoking rates between countries, with 80% of smokers living in low and middle-income countries (LMICs) (US NCI and WHO 2016). Because substance use is dynamic across time and place, evaluation research needs to be ongoing and as localised as resources permit. As Babor and colleagues note (2010b), communities often want a local evidence base to underpin interventions and distrust data from another place (or country). This has special resonance in rural townships, as discussed in Chapter 4.

It is not surprising that the most robust systems for monitoring substance use are found in high-income countries (Hall 2018; WHO 2017b). These countries can afford to fund evaluation research, both directly and indirectly (e.g. through the financial support they are able to provide to their university sectors). Resourcing, or lack of it, is an important reason why low-income countries are less likely to have well-developed policy platforms and the research infrastructure to evaluate the effectiveness of their policies (see WHO 2011).

Figure 1.1 illustrates some interrelationships between evaluation research and policy formation at national, regional and global levels.

Before discussing key issues relating to **data generation** and **research design**, in broad terms Figure 1.1 indicates that evaluation research can influence policy and that changes in policy can change the foci of evaluation research. National, regional and global policy platforms can all

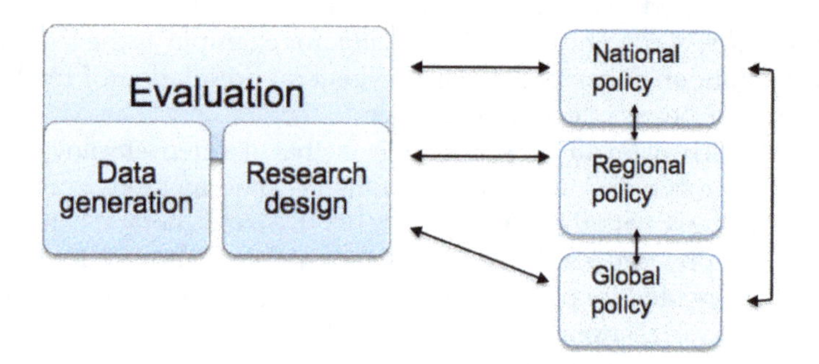

Figure 1.1 Interrelationships Between Evaluation Research and Substance Use Policy.

influence each other as well. For instance, a national policy that is considered successful may influence global policy. Global policy plays a role in shaping aspects of national policies, perhaps best evidenced by the influence of UN treaties on illicit drugs upon domestic criminal laws in signatory countries.

Evaluation research within countries is collated by regional and global agencies to map international patterns. For example, the UNODC (2016a; 2016b) collects information from countries about the prevalence of drug consumption in survey data, the number and weight of drug seizures, the sizes of coca crops eradicated, expert opinion and so on. The WHO similarly collates survey data on alcohol consumption, alcohol sales, alcohol-related traffic accidents and other alcohol-related mortality (WHO 2010, 2014). The WHO (2017b) also gathers data on the prevalence of tobacco consumption and its harms. In these ways, national evaluation research provides information relevant to all tiers of policy.

Regional and global agencies are active in evaluation research. They help set standards for valid and effective evaluation research. In some cases, they may facilitate research training in low-income countries.

1.3.1 Data Generation

Data on psychoactive substance use come from many sources. **Survey data** are generated by researchers when they interview a sample of participants about their substance use. Surveys have arguably been the keystone of evaluation research on policies towards psychoactive substances. They have several strengths. When collecting data under conditions that assure confidentiality on samples of people who are representative of the general population, they may be a reliable indicator of the proportion of a population who consume these substances.

Surveys also give good indications about the frequency *with* which individuals consume different substances and how they consume them. This information is essential to understand the percentage of the community who may be at risk of the types of harm associated with intoxication, different routes of administration (e.g. injecting drug use), and regular use and dependence (see above, 1.1.1). Some surveys, such as the Global Drug Survey,[3] do not attempt to secure representative samples for prevalence research, but concentrate on collating data on regular users of illicit drugs in order to better understand substance use behaviours and self-reported harms arising from such use.

Surveys can measure the nature and extent of poly-substance use, such as the concurrent use of alcohol, tobacco and one or more illicit drugs. Survey data also enables researchers to examine the strength of relationships between substance use and age, gender, health, mental health, criminal behaviour, housing stability, employment, education, family relationships and so forth. Participants can also be asked about their future intentions, such as whether a tax excise will influence their tobacco smoking. Surveys can also be used to assess substance use in particular groups, such as young people in secondary schools, people who inject drugs, people who have been arrested, or people who are imprisoned.

Procedures employed to gather survey data (e.g. face-to-face, telephone, online) vary in the resources and time required. The appropriateness of procedures depends on the research questions under examination. Studies that aim to estimate the prevalence of substance use in the general population require large sample sizes for statistical confidence and so are expensive and time consuming. For example, for a country with a population of approximately 25 million people, Australia's lead survey on substance use produces data every three years on a sample of about 25,000 participants (AIHW 2017). The lead US survey annually interviews close to 70,000 participants (SAMHSA 2017).

Surveys are subject to selection effects, or selection bias. Selection effects work to undermine the representativeness of the sample of participants in a survey. These effects occur when, among other things, certain groups are: difficult to reach (e.g. people who do not have a home or access digital technology); or are reluctant to talk about their substance use behaviours (Harrison and Hughes 1997).

Qualitative data are critical in understanding substance use. They are frequently collected in larger surveys of substance use in the form of free text from participants about their substance use. Incorporating qualitative data into surveys in this way is common across the social sciences. Experts' opinions on psychoactive substances may be described as a form of qualitative data. The term 'expert' includes professionals from fields of social welfare, health, law enforcement, customs, licensing regulation, advertising regulation, secondary education and the like. Their opinions might also be described as 'intelligence' or 'evidence'. They may be valuable in estimating prevalence of use in the absence of quantitative data.

Experts assist in understanding the harm-potential of substances, as is the case in the Global Burden of Disease projects discussed earlier. Naturally, experts' views on the relative efficacy of strategies are important but may be limited. People who use drugs regularly may also be able to provide an informed and useful opinion on, for instance, (a) the effects of strategies on them or (b) activity within a drug market – as indicated by drug price, purity, the appearance of new drugs and the ease with which purchases can be made.

Official statistics, sometimes called **event data** (Babor et al. 2010b), are generated by agencies whose activities are related to psychoactive substances. When we recall the discussion in Section 1.2 of the breadth of agencies that may play a role in substance use policy, event data may take multiple forms.

Event data from law enforcement agencies, customs agencies, courts and forensic laboratories may include: numbers of drug-related arrests; seizures of drugs and black-market alcohol or tobacco; prosecutions for breaches of alcohol or tobacco licensing regulations; and crop eradication activities.

Health agencies produce event data on *mortality* whenever they, for instance, deal with fatal opioid overdoses or car accidents. The wide array of situations in which health agencies respond to diseases caused by or associated with substance use produce data on drug-related *morbidity.* Other examples of valuable event data are treatments rates for substance use dependence, demand for needle exchange programs, prescription data and so on (Ministerial Council on Drug Strategy 2011).

A major limitation of event data is that there can be significant time lags between when event data are collected, collated, analysed and released (Prichard et al. 2017). Interpreting event data can also be challenging. An agency's activities may change because of fluctuations (up or down) in funding, or because of policy changes. Consequently, it may be difficult to determine whether changes in event data reflect changes in the activities of agencies, or, changes in substance use. Does an increase in prescriptions for buprenorphine, for example, reflect a rise in heroin dependence, or an increase in the funding of the treatment? Do decreases in drug-related arrests indicate a downward trend in the prevalence of drug consumption, or a change in policing strategies? Should we interpret decreases in drug seizures as evidence of the effectiveness of interdiction strategies, or a change in supply routes? These factors complicate the interpretation of event data in evaluation research (e.g. Willis, Anderson and Homel 2011).

Industry data comprise statistics generated by legal commercial markets for alcohol and tobacco, such as sales and taxation data. When available, sales data give a measure of the volume and rate of consumption – typically at the national level (Babor et al. 2010a; USNCI and WHO 2016). Generally, sales data do not provide a clear picture on frequency of consumption because of the way that they are collated by industry. There is also often a delay between the purchase and consumption of alcohol and tobacco. In addition, commercial sales data do not count non-commercial consumption. This is estimated to account for about 8% of the global tobacco market (USNCI and WHO 2016). Non-commercial consumption of alcohol – that includes home-brewed beverages as well as black-market products – is estimated to account for 30% of the global market (Rehm et al. 2014).

Table 1.1 illustrates categories of data on substance use at local, national and international scales. It draws on examples from one-off studies as

Table 1.1 Examples of Data Sourced in Single Studies and Monitoring Systems (Local, National and International)

	Local	National	International
Surveys	Interrelationships between injecting drug use & incarceration in Baltimore were studied with 3,245 interviews (Genberg et al. 2015)	US *National Survey on Drug Use and Health* (SAMHSA 2017)	Worldwide adult & youth tobacco surveys (WHO 2017b) *Global Drug Survey* (e.g. Labhart et al. 2017)
Qualitative evidence	Following WHO guidelines, ethnographic & qualitative research techniques were used to gather data on injecting drug use behaviours in Tehran (Razzaghi et al. 2006)	*Ecstasy and Related Drug Reporting System* (Uporova et al. 2018) consults with professionals & indirect sources about ecstasy and 'party' drugs	*Global Burden of Disease* (Degenhardt et al. 2013) consults with experts in HIV & drugs
Event data	BOCSAR monitors policing incidents involving methamphetamines in 28 local government areas of New South Wales, Australia*	*EMCDDA UK Drug Report 2017* (EMCDDA 2017) incorporates drug-related arrest & medical emergencies	*Joint Airport Interdiction Taskforces* (UNODC 2016a, 2016b) monitors airport drug seizures by number & volume
Industry data	Health Canada reports annual sales of tobacco products in provinces and territories**	NHS Health Scotland reports alcohol consumption & price as part of *Monitoring & Evaluating Scotland's Alcohol Strategy*[+]	*Global Information System on Alcohol and Health* (WHO 2014) collates publicly available data from economic operators & industry-supported organisations

* Bureau of Crime Statistics and Research. Accessed 6 April 2018. www.bocsar.nsw.gov.au/Pages/bocsar_pages/Amphetamines.aspx.

** Health Canada. National and provincial/territorial tobacco sales data 2017 – Page 4. Accessed 6 April 2018. www.canada.ca/en/health-canada/services/publications/healthy-living/federal-provincial-territorial-tobacco-sales-data/page-4.html.

[+] NHS Health Scotland. MESAS alcohol sales and price update May 2016. Accessed 6 April 2018. www.healthscotland.scot/publications/mesas-alcohol-sales-and-price-update-may-2016.

well as ongoing monitoring systems. 'Local' is a necessarily broad term referring to subnational contexts, such as schools, townships, cities, provinces, territories or states. The boundaries between local and national data blur inasmuch as national data collection can be disaggregated (sample sizes permitting) to present data on drug use in provinces or regions of a country.

1.3.2 Research Designs

Data on substance use can be analysed with a wide assortment of qualitative and quantitative research techniques, details of which are not discussed here. A broader question of more importance is how techniques for data analysis can be incorporated into the design of evaluation studies.

Descriptive statistics are an essential starting point for evaluation. It is through monitoring tobacco consumption, for instance, that the WHO has determined that smoking rates are declining in high-income countries while increasing in LMICs (USNCI and WHO 2016). Likewise, some agencies use downward trends in drug-related harms as indications of the success of their strategies (e.g. Ministerial Council on Drug Strategy 2011). The value of monitoring descriptive data will be discussed further in later chapters of this book.

But as discussed above, the fact that substance use can change over time and differs between areas for a variety of reasons means that descriptive statistics on time periods or regions may reveal little about the effects of particular strategies or policies. How can a strategy's effect (if any) be isolated from the influence of multiple other potential factors?

Randomised controlled trials (RCT) are arguably the best empirical method for controlling the effects of extraneous factors but their application is limited for practical reasons to evaluating comparatively small-scale interventions. For instance, RCTs have been conducted to compare different treatments for groups of high-frequency drug users, to evaluate the effectiveness of drug-prevention programs in school settings (Babor et al. 2010b), or to compare criminal law interventions for drink-driving and drug-driving offenders (Sherman et al. 2015). However, even at the national level within high-income countries it is rare for governments to implement drug strategies in ways that make them amenable to RCTs (Hall 2018); RCTs are also expensive to implement and involve ethical complexities (Babor et al. 2010b), such as the fact that control groups may not be able to access a potentially beneficial treatment. RCTs are not an option for comparing the effectiveness of strategies between countries.

Opportunities for **quasi-experimental studies** arise when governments change policy in some way (including terminating a strategy), or when other circumstances (e.g. a sudden and sustained drug shortage) affect markets for psychoactive substances (see Degenhardt et al. 2005). Unlike RCTs, quasi-experiments, also called natural experiments (Babor et al. 2010a, 2010b), do not involve random allocation of groups to control and experimental conditions. This means that they have less capacity to control for extraneous factors, such as selection effects or placebo effects. Nonetheless, well-designed and executed quasi-experimental studies can provide robust findings for policy evaluation and causal inference (see Babor et al. 2010b).

Frequently, analyses of the effects of changing strategies or circumstances have to be conducted retrospectively using data collected for other purposes that may not be ideal for the evaluation task (Hall 2018). In the context of changes to government strategies, greater confidence can be achieved if studies are designed and implemented in advance of a change, especially when a range of data can be collected during a baseline period before the policy change is affected (Hall 2018). If sufficient trend data are available it may be possible to conduct time series analyses (Babor et al. 2010b). Studies of more than one 'site' (e.g. school, town or country) broadly matched on demographic and economic profiles may enable comparisons of experimental sites (where a change has occurred) and control sites (where no change has occurred).

1.4 Conclusion

This chapter serves as an introduction to the remainder of the book, and so provides an overview of the harms of substance use, the types of strategies that are used to reduce those harms, and the ways in which these strategies are evaluated. Great diversity is found in the nature of psychoactive substances, their effects and the markets that support them. Likewise, there is considerable diversity in (a) strategies to counter harm and (b) measurement methods. Chapter 1 did not discuss in detail specific problems associated with measurement at the international, national or local level. These tasks will be picked up respectively by Chapters 3, 4 and 5.

Chapter 2 introduces a new measurement method, wastewater analysis (WWA). We seek to avoid two main risks in approaching the topic of WWA for the first time: excess optimism about its capabilities and pessimistic indifference to yet another way of measuring substance use. Chapter 2 argues that WWA deserves careful consideration. It is not a 'silver bullet' that replaces existing approaches; it complements traditional approaches to studying substance use because it measures substance use in a very different way to other methods and its empirical strengths provide a counterweight to some of the shortcomings of traditional research and evaluation methods.

Notes

1. Major classes of psychoactive substances include: opioids, cannabis, cocaine, amphetamine-type substances (e.g. amphetamine, methamphetamine, methylenedioxymethamphetamine, or 'ecstasy'), sedatives (e.g. benzodiazepines) and hallucinogens (e.g. lysergic acid diethylamide, phencyclidine, mescaline).
2. See National Survey on Drug Use and Health. Accessed 6 April 2018. https://nsduhweb.rti.org/respweb/homepage.cfm.
3. Global Drug Survey. Accessed 5 April 2018. www.globaldrugsurvey.com/.

References

Australian Institute of Health and Welfare. 2017. *National drug strategy household survey 2016: Detailed findings.* Drug Statistics series no. 31. Cat. no. PHE 214. Canberra: Australian Institute of Health and Welfare.

Babor, T. F., R. Caetano, S. Casswell et al. 2010a. *Alcohol: No ordinary commodity: Research and public policy*. Oxford: Oxford Univ. Press.

Babor, T. F., J. P. Caulkins, G. Edwards et al. 2010b. *Drug policy and the public good.* Oxford: Oxford Univ. Press.

Brewer, A. J., C. J. Banta-Green, C. Ort, A. E. Robel, and J. Field. 2016. Wastewater testing compared with random urinalyses for the surveillance of illicit drug use in prisons. *Drug and Alcohol Review* 35:133–7.

Bureau of Crime Statistics and Research. Amphetamines statistics and Cocaine statistics for NSW. Accessed 6 April 2018. www.bocsar.nsw.gov.au/Pages/bocsar_pages/Amphetamines.aspx

Campbell, D.T. 1969. Reforms as experiments. *American Psychologist*, 24(4):409–29.

Castiglioni, S., and L. Vandam. 2016. A global overview of wastewater-based epidemiology. In *Assessing illicit drugs in wastewater: Advances in wastewater-based drug epidemiology*, ed. S. Castiglioni, 45–54. Luxembourg: Publications Office of the European Union.

Castiglioni, S., L. Vandam, and P. Griffiths. 2016. Conclusions and final remarks. In *Assessing illicit drugs in wastewater: Advances in wastewater-based drug epidemiology*, ed. S. Castiglioni, 75–76. Luxembourg: Publications Office of the European Union.

Collins, D., and H. M. Lapsley. 2008. *The costs of tobacco, alcohol and illicit drug abuse to Australian society in 2004/05*, National Drug Strategy Monograph Series No. 64. Canberra: Department of Health and Ageing.

Commonwealth of Australia (Department of Health). 2017. National Drug Strategy 2017–2026. www.health.gov.au/sites/default/files/national-drug-strategy-2017-2026_1.pdf

Degenhardt, L., F. Charlson, B. Mathers et al. 2014. The global epidemiology and burden of opioid dependence: Results from the global burden of disease 2010 study. *Addiction* 109:1320–33.

Degenhardt, L., P. Reuter, L. Collins, and W. Hall. 2005. Evaluating explanations of the Australian 'heroin shortage'. *Addiction* 100:459–69.

Degenhardt, L., H. A. Whiteford, A. J. Ferrari et al. 2013. Global burden of disease attributable to illicit drug use and dependence: Findings from the Global Burden of Disease Study 2010. *The Lancet* 382:1564–74.

European Monitoring Centre for Drugs and Drug Addiction. 2015. The EU drugs strategy (2013–20) and its action plan (2013–16). www.emcdda.europa.eu/attachements.cfm/att_212356_EN_EMCDDA_POD_2013_New%20EU%20drugs%20strategy.pdf

European Monitoring Centre for Drugs and Drug Addiction. 2017. *United Kingdom, Country drug report 2017*. Luxembourg: Publications Office of the European Union.

Genberg, B. L., J. Astemborski, D. Vlahov, G. D. Kirk, and S. H. Mehta. 2015. Incarceration and injection drug use in Baltimore, Maryland. *Addiction* 110:1152–9.

Hall, W. 2018. The future of the international drug control system and national drug prohibitions. *Addiction* 113:1210–23.

Hall, W. D., R. Room, S. Bondy. 1999. Comparing the health and psychological risks of alcohol, cannabis, nicotine and opiate use. In *The health effects of cannabis*, ed. H. Kalant, W. Corrigall, W. Hall, and R. Smart, 475–506. Toronto: Addiction Research Foundation.

Harrison, L., and A. Hughes. 1997. *The validity of self-reported drug use: Improving the accuracy of survey estimates*. Rockville, MD: U.S. Department of Health and Human Services, National Institutes of Health, National Institute on Drug Abuse, Division of Epidemiology and Prevention Research

Henry, S. 2012. Expanding our thinking on theorizing criminology and criminal justice? The place of evolutionary perspectives in integrative criminological theory. *Journal of Theoretical and Philosophical Criminology* 4:62–89.

Labhart, F., J. Ferris, A. Winstock, and E. Kuntsche. 2017. The country-level effects of drinking, heavy drinking and drink prices on pre-drinking: An international comparison of 25 countries. *Drug Alcohol Review* 36:742–50.

Ledford, H. 2015. How to solve the world's biggest problems. *Nature* 525:308–11.

Lim, S. S., T. Vos, A. D. Flaxman et al. 2012. A comparative risk assessment of burden of disease and injury attributable to 67 risk factors and risk factor clusters in 21 regions, 1990–2010: A systematic analysis for the Global Burden of Disease Study 2010. *The Lancet* 380:2224–60.

Loxley, W., J. W. Toumbourou, T. Stockwell et al. 2004. *The prevention of substance use, risk and harm in Australia: A review of the evidence*. Canberra: The National Drug Research Centre and the Centre for Adolescent Health.

Ministerial Council on Drug Strategy. 2011. *National drug strategy 2010–2015: A framework for action on alcohol, tobacco and other drugs*. Canberra: Ministerial Council on Drug Strategy.

Murray, C. J. L., T. Vos, R. Lozano et al. 2012. Disability-adjusted life years (DALYs) for 291 diseases and injuries in 21 regions, 1990–2010: A systematic analysis for the Global Burden of Disease Study 2010. *The Lancet* 380:2197–223.

National Cancer Institute and Centers for Disease Control and Prevention. 2014. *Smokeless tobacco and public health: a global perspective*. NIH publication no. 14–7983. Bethesda, MD: U.S. Department of Health and Human Services, Centers for Disease Control and Prevention, and National Institutes of Health, National Cancer Institute Available from: http://cancercontrol.cancer.gov/brp/tcrb/global-perspective/SmokelessTobaccoAndPublicHealth.pdf

Nutt, D., L. A. King, W. Saulsbury, and C. Blakemore. 2007. Development of a rational scale to assess the harm of drugs of potential misuse. *The Lancet* 369:1047–53.

Office of National Drug Control Policy. 2004. *The economic costs of drug abuse in the United States, 1992–2002*. Washington, DC: Executive Office of the President.

Prichard, J., F. Y. Lai, E. van Dyken et al. 2017. Wastewater analysis for estimating substance use: Implications for law, policy and research. *Journal of Law and Medicine* 24:837–849.

Razzaghi, E. M., A. R. Movaghar, T. C. Green, and K. Khoshnood. 2006. Profiles of risk: A qualitative study of injecting drug users in Tehran, Iran. *Harm Reduction Journal* 3:12.

Rehm, J., D. Baliunas, S. Brochu et al. 2006. *The costs of substance abuse in Canada 2002*. Ottawa: Canadian Centre on Substance Abuse.

Rehm, J., S. Kailasapillai, E. Larsen et al. 2014. A systematic review of the epidemiology of unrecorded alcohol consumption and the chemical composition of unrecorded alcohol. *Addiction* 109: 880–93.

Schifano, F., L. Orsolini, D. Papanti, and J. Corkery. 2016. NPS: Medical consequences associated with their intake. In *Neuropharmacology of New Psychoactive Substances (NPS)*, ed. M. Baumann, R. Glennon, and J. Wiley, 351–380. Cham: Springer.

Sherman, L. W., H. Strang, G. Barnes et al. 2015. Twelve experiments in restorative justice: The Jerry Lee program of randomized trials of restorative justice conferences. *Journal of Experimental Criminology* 11:501–40.

Substance Abuse and Mental Health Services Administration. 2017. *Key substance use and mental health indicators in the United States: Results from the 2016 National Survey on Drug Use and Health* (HHS Publication No. SMA 17–5044, NSDUH Series H-52). Rockville, MD: Center for Behavioral Health Statistics and Quality, Substance Abuse and Mental Health Services Administration. www.samhsa.gov/data/sites/default/files/NSDUH-FFR1-2016/NSDUH-FFR1-2016.htm.

United Nations Convention against Illicit Traffic in Narcotic Drugs and Psychotropic Substances. 1988. www.unodc.org/unodc/en/treaties/illicit-trafficking.html

United Nations Convention on Psychotropic Substances. 1971. www.unodc.org/unodc/en/treaties/psychotropics.html

United Nations Office on Drugs and Crime. 2016a. *World drug report 2016*. New York: United Nations.

United Nations Office on Drugs and Crime. 2016b. *UNODC annual report: Covering activities during 2016*. Vienna: UNODC www.unodc.org/documents/AnnualReport2016/2016_UNODC_Annual_Report.pdf.

United Nations Office on Drugs and Crime. 2017. *World drug report 2017 (Booklet 5 – The drug problem and organized crime, Illicit financial flows, corruption and terrorism)*. New York: United Nations.

United Nations Single Convention on Narcotic Drugs. 1961. www.unodc.org/unodc/en/treaties/single-convention.html

United States National Cancer Institute, and World Health Organization. 2016. *The economics of tobacco and tobacco control*. Tobacco Control Monograph 21. National Cancer Institute http://cancercontrol.cancer.gov/brp/tcrb/monographs/21/index.html.

Uporova, J., A. Karlsson, R. Sutherland, and L. Burns. 2018. *Australian trends in ecstasy and related drug markets 2017. Findings from the Ecstasy and Related Drugs Reporting System (EDRS)*. Sydney: National Drug and Alcohol Research Centre, University of New South Wales.

Willis, K., J. Anderson, and P. Homel. 2011. Measuring the effectiveness of drug law enforcement. *Trends and Issues in Crime and Criminal Justice* 406:1–7.
World Health Organization. 2003. *WHO Framework Convention on Tobacco Control.* www.who.int/fctc/text_download/en/.
World Health Organization. 2010. *Global strategy to reduce the harmful use of alcohol.* Geneva: World Health Organization.
World Health Organization. 2011. *Global Adult Tobacco Survey (GATS): Implementing agency selection guidelines.* Version 3.0. www.who.int/tobacco/surveillance/3_GATS_ImplementingAgencySelectionGuidelines_v3_FINAL_09Dec2011.pdf?ua=1.
World Health Organization. 2012. *European action plan to reduce the harmful use of alcohol 2012–2020.* Denmark: WHO Regional Office for Europe.
World Health Organization. 2014. *Global status report on alcohol and health 2014.* Geneva: World Health Organization.
World Health Organization. 2017a. *World health statistics 2017: Monitoring health for the SDGs, Sustainable Development Goals.* Geneva: World Health Organization.
World Health Organization. 2017b. *WHO report on the global tobacco epidemic 2017: Monitoring tobacco use and prevention policies.* Geneva: World Health Organization.
World Health Organization. 2017c. *Technical Briefing: Alcohol control interventions for Appendix 3 of the Global Action Plan for Non Communicable Diseases.* www.who.int/ncds/governance/harmful_use_of_alcohol.pdf?ua=1

chapter two

Understanding Wastewater Analysis

How It Works; Its Strengths and Limitations

INTRODUCTION

Chapter 1 overviewed what could be described as the global problem of substance use and the main methods that have traditionally been used to monitor it. This second chapter explains the essential features of wastewater analysis (WWA) for measurement of substance consumption. It examines how, within the space of a decade, the efficacy of the WWA method has been established so that, from a scientific perspective, the field has moved beyond a 'proof of concept' into 'application'. This chapter is designed to be accessible to readers with a social science background. Its primary objective is to equip readers with an accurate understanding of how WWA works and its main strengths and limitations as an additional tool for monitoring substance use, particularly illicit drug consumption. This chapter may also be useful for non-scientists who are interested in initiating new WWA (also called 'wastewater-based epidemiology') research within their country.

A recurring theme in this chapter is the interdisciplinary nature of WWA, which involves academic and non-academic experts. Scientists – analytical chemists, pharmacologists and sewer engineers – are at the engine room of data production, analysis and interpretation. Experts in health and criminal justice portfolios add value to WWA data by (a) designing projects to best address policy-relevant issues and (b) interpreting results. It seems likely that, as argued elsewhere (Prichard et al. 2017), increased engagement by policy makers in these portfolios is imperative if WWA is to realise its full potential in making and evaluating drug policy.

Chapter 2 has five sections. Section 2.1 explains the origins of WWA in environmental science, which studies the fate of chemicals in sewage water and aquatic ecosystems. Section 2.2 provides an explanation as to how WWA works – considering features of the human metabolic system, how wastewater is sampled, and key steps in laboratory analyses. This section describes some key uncertainties in WWA data. An overview of the ethical issues relating to WWA research is supplied in 2.3. The current main applications of WWA are described in 2.4, including population monitoring by the European Monitoring Centre for Drugs and Drug Addiction. The chapter concludes in 2.5 with a discussion of the strengths and weaknesses of WWA in comparison with established methods.

2.1 Origins of Wastewater Analysis

The antecedents of WWA can be dated to the 1970s when environmental researchers examined the capacity of sewage treatment plants (WWTPs) to prevent chemicals from entering the environment. The chemicals of concern included those *disposed* of by people down kitchen and bathroom sinks, through washing machines and so forth (e.g. Ahel and Giger 1985; Ternes 1998). But attention also turned to chemicals that were *excreted* by people in toilets after they had consumed medicine (e.g. Osemwengie and Steinberg 2001; Garrison, Pope and Allen 1976) and substances such as caffeine and tobacco (Hignite and Azarnoff 1977). The field was primarily interested in the fate of chemicals in the environment and their subsequent potential negative effects on human health (e.g. Daughton 2003) and aquatic organisms, including fish (Jones-Lepp et al. 2004).

Christian Daughton was one of the first to recognise that the techniques developed by chemists to measure pharmaceuticals in sewage water could be adapted to map population-level consumption of illicit drugs in time and space. In retrospect, Daughton's short 2001 publication showed remarkable farsightedness. He appreciated that he had identified a "rare bridge between the environmental and social sciences" (2001, 348). While highlighting some shortcomings of WWA – including its inability to determine numbers of users – Daughton accurately identified several key advantages of WWA. These included the breadth of the data, their objectivity, the capacity for data to be generated on a very wide variety of substances in near real-time, and the fact that WWA was non-intrusive and did not identify individual users.

Although Daughton arguably understated the value of self-report methods, his analysis of the comparative advantages of WWA has proven correct. He contended that WWA would circumvent certain problems affecting surveys, such as cost, selection effects, sampling errors, delays in reporting and inaccurate reporting (see Daughton 2001, 358). Daughton expected WWA to be of interest at the national and local level to both health and criminal justice agencies. He also anticipated that WWA could be useful for schools, thereby foreshadowing the study of drug use in particular buildings.

The most practical aspect of Daughton's publication was his proposal for 'back-calculating' population-level drug consumption. This involved combining data from laboratory tests of sewage water with information about human metabolism (pharmacokinetics), how long chemicals persist in sewage water and the size of the population within the catchment area of a WWTP.

North American researchers were the first to act upon Daughton's ideas. It appears that the first study of an illicit drug, methamphetamine, was conducted in California and was reported by Stuart Khan and Jerry Ongerth in Minneapolis, 2003, at a conference of the National Ground Water Association (Jones-Lepp et al. 2004). The first published study shortly followed, led by a former colleague of Daughton, Tammy Jones-Lepp. This study quantified methamphetamine and ecstasy (MDMA) and four pharmaceuticals in samples taken from South Carolina, Nevada and Utah. The authors made reference to the implications of their findings for law enforcement agencies and expressed a hope that the findings might "spur socio-economic researchers to [use WWA] to better target antidrug campaigns" (Jones-Lepp et al. 2004, 438).

The article that marked the beginning of systematic research activity on WWA was published by an Italian team that included Ettore Zuccato, Sara Castiglioni and the late Davide Calamari. Zuccato et al. (2005) quantified levels of cocaine in samples taken from the River Po and urban WWTPs. They applied the back-calculation methods that Daughton (2001) had suggested. The findings indicated significantly higher levels of cocaine consumption than estimated by Italian agencies. In a media statement Zuccato pointed out that if his team's estimates were accurate, the amount of cocaine consumed in the catchments was 80 times higher than other data sources indicated (Bhattacharya 2005).

Not surprisingly, the study generated considerable media interest. The Italian tabloid, *la Repubblica.it* (2005), was among the first outlets to report the findings. News pieces were published in the *New Scientist* (Bhattacharya 2005) and *Chemistry World* (Down 2005). Within days, stories appeared in mainstream English-speaking media outlets around the world (e.g. *The Guardian* – Radford 2005; *BBC News* 2005; *The Scotsman* 2005; *Australian Broadcasting Corporation* 2005; *Houston Chronicle* 2005) – sometimes with sensational headlines, such as 'A river of cocaine' (*The Mirror* 2005) and 'Shock over cocaine river residues' (*CNN.com* 2005).

There is no doubt that media coverage, which has continued to the present day (e.g. Prichard et al. 2018), has benefitted the WWA field by, for example, encouraging researchers globally to investigate the potential of these scientific techniques. However, the interest that the media has demonstrated in WWA itself carries implications for human research ethics that are discussed further below.

Before moving on to catalogue the main developments in WWA and its current capabilities, it is worth pausing to note regional differences in research activity pre- and post-2005. Europe has been the centre of WWA activity since 2005 as indicated by the numbers of WWA researchers and the quality and quantity of material published by European research teams. Other significant research groups have emerged in Australia and North America.

Important research has been conducted in the USA (e.g. Hart and Halden 2020; Chiaia, Banta-Green and Field 2008; Banta-Green et al., 2009; Banta-Green and Field, 2011; Banta-Green et al. 2016; Brewer et al. 2016) but the level of research investment in the USA has not matched that of Europe. This is puzzling given (a) the size of American research capability, (b) the fact that early WWA projects were American (Jones-Lepp et al. 2004) and (c) that the concept was proposed by a senior member of the EPA (Daughton 2001), a large Federal agency with a mandate to protect human health – the parameters of which complement epidemiology.

However, it is feasible that some WWA in the USA has been conducted by public agencies without publication and without the involvement of the university sector. In 2006 the *Washington Post* reported the contents of documents publicly released by local government in Fairfax County, Virginia (Turque 2006). The memo noted that during the administration of President George W. Bush local government authorities had granted permission to the White House Office of National Drug Control Policy (ONDCP) to sample sewage from the County's WWTPs with a view to estimate cocaine consumption with WWA. This formed part of a broader assessment by the ONDCP of the utility of WWA to monitor drug consumption across several communities, which were not identified. The ONDCP arranged sampling over a five-day period. Laboratory analyses were conducted by the Armed Forces Institute of Pathology in Maryland.

The *San Diego Reader* reported similar details, but indicated the ONDCP project had taken 160 samples from 34 sites (Potter 2006). The results of the analyses do not appear to have been released. It is unclear whether authorities preferred not to disclose the findings because of concerns about criticisms on civil liberties grounds, as suggested by Potter (2006). It is also not known whether any American authorities have continued to conduct WWA.

2.2 How Does Wastewater Analysis Work?

In any empirical research it is critical for scholars to *know* their data and especially the limitations in regards to their interpretation. This applies across all the sciences and social sciences. It is true of quantitative and qualitative data including analyses of text – criminal cases, policy documents, narratives from participants and so forth. How are WWA data derived and what do they represent? Although there are multiple ways in which WWA can be improved, it is accepted as an efficacious and robust approach to study human behaviour. In this book our emphasis is on WWA's capacity to study population level consumption of major illicit drugs of concern: cannabis, methamphetamine, cocaine, MDMA, methadone, LSD and ketamine (Mastroianni et al. 2017). WWA can equally be used to study (a) consumption of many pharmaceuticals (Lai et al. 2011), alcohol (Reid et al. 2011) and tobacco (Castiglioni et al. 2015); (b) communal health (Ryu et al. 2016a); and (c) levels of human exposure to pollutants (Rousis, Zuccato and Castiglioni 2016).

The central aspects of WWA are simple enough. Chemicals arising from illicit drug consumption are excreted by the human body into sewerage systems. Sewage water is sampled (usually over a 24 period), typically at the inflow to sewage treatment plants (WWTPs). Then the samples are analysed at a laboratory and the results are used to estimate consumption of illicit drugs in the population contributing to the wastewater treatment plant. An overview of the process is depicted Figure 2.1.

2.2.1 *Chemical Excretion from the Human Body into Sewers*

The human body has metabolic and other processes for transforming and excreting foreign chemicals that enter the body either deliberately, e.g. by eating, drinking, smoking, or through unintended exposure to pollutants in our environment. The processes are complex and varied, not least because the chemicals the body needs to respond to number in the thousands (EMCDDA 2016).

Illicit drugs, alcohol and tobacco exert their effect upon the human body once they enter the bloodstream and can interact with their particular target, such as the central nervous system. The body is quite a hostile environment to many drugs, as it is for many other substances that are consumed and even some substances that are produced by the body itself. Enzymes within organs such as the liver are the main source of hostility and they convert drugs into other chemicals (called metabolites) that are easy for the body to eliminate via urine, faeces, etc. It is not the case that all drugs are metabolized into one metabolite and it is not the case that all drugs are completely metabolized within the body – some drugs produce multiple metabolites, some are completely metabolized and some are barely metabolized at all and almost all of the dose consumed is eliminated unchanged (EMCDDA 2016). The substances excreted by the body are broadly consistent regardless of the route of administration. In other words, the *products* of excretion do not differ if a psychoactive substance is swallowed, smoked, injected or taken as a suppository (Prichard et al.

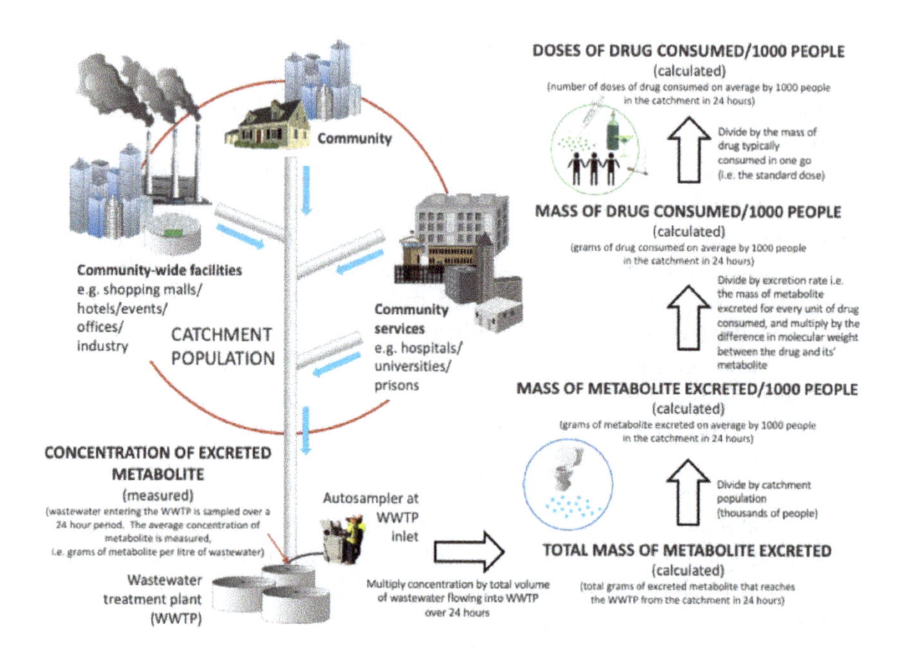

Used with permission from the Queensland Alliance for Environmental Health Sciences. The symbols/images used in Figure 1 in the report were provided courtesy of the Integration and Application Network, University of Maryland, Center for Environmental Science (ian.umces.edu/symbols/).

Figure 2.1 Generic Flow Chart of WWA Process from Sources of Wastewater to Analysis.

2012). With the exception of cannabis, most psychoactive substances are excreted within a few hours of use (Baselt 2008). Metabolic rates can differ markedly between individuals but in the large groups of people who contribute to WWTPs it is reasonable to use average excretion rates on the assumption that the variability in the populations' metabolic systems is lower than that of the individuals who comprise it (Prichard et al. 2012).

In order to measure population consumption of particular substances, WWA scientists first need to select the best chemical indicators of human excretion, or 'biomarker'. The processes involved in selecting biomarkers are complex and research continues on improving our understanding of the properties of WWA biomarkers. The most suitable biomarkers remain stable in wastewater and exist in sufficient quantities to be easily detected (see Castiglioni, Bijlsma, Covaci et al. 2016). Ideally, biomarkers should be uniquely related to human excretion rather than direct disposal of substances (EMCDDA 2016). This topic is discussed further below in relation to illicit drugs.

The human body also excretes chemicals as by-products of the digestive system and the immune system. Consequently, researchers have been able to use some biomarkers to study aspects of diet, such as the consumption of caffeine (O'Brien et al. 2014). Biomarkers for numerous diseases have been studied too, including multiple forms of cancer (Daughton 2018). Other biomarkers can act as indicators of oxidative stress (Ryu et al. 2016a) and human exposure to pollutants, such as pesticides and phthalates (e.g. Gonzalez-Marino et al. 2017).

2.2.2 Sampling Wastewater

So much of the initial focus in WWA fell upon laboratory techniques (addressed in 2.2.3) that researchers largely overlooked a relatively simple question: are sample-collection techniques valid? In 2010 a Swiss sewer engineer, Christoph Ort, and colleagues published two papers examining this issue. The papers had an immediate and lasting effect on the field and have provided practical ways for teams to improve their sampling methods and reporting of results.

Ort, Lawrence, Reungoat et al. (2010) calculated the margins of error – or 'uncertainties' – that are applicable to WWA results as a result of sampling. Their statistical modelling indicated that the uncertainties introduced by sampling could be far greater than the uncertainties arising from chemical analysis. In other words, more error could be introduced by the way in which samples of sewage water were collected and stored than through the techniques used to analyse the samples in the laboratory.

The kernel of the problems Ort and colleagues underscored was unrepresentative or biased sampling. By way of analogy, an extended social science survey of individuals on the streets of a large city would experience 'pulses' of people at rush-hours. Compared with people on the streets during the day, the rush-hour participants might have different socio-economic backgrounds in terms of employment, age, health and the like. Therefore surveying only at rush hours, or only between rush hours, would bias samples and thereby reduce a study's representativeness.

Similarly, Ort, Lawrence, Reungoat et al. (2010, 6289) explained that:

> ... wastewater may appear as a continuous stream, but it is actually composed of a number of intermittently discharged, individual wastewater packets from household appliances, industries, or subcatchments in pressurized sewer systems. The resulting heterogeneity can cause significant short-term variations of pollutant loads.

Sampling protocols in WWA studies have to carefully account for the specific conditions of the sewerage system from which the samples are drawn. It is not likely to be the case that the individuals who consume drugs contribute to the wastewater stream in a continuous fashion; there are likely to be peaks and troughs over time in the amount of drugs and metabolites present in the stream. Without appropriate sampling procedures the representativeness of WWA results can be significantly undermined. For example, if a wastewater sample is collected at the moment that a 'packet' containing a lot of drug/metabolite is passing through the sewer then an overestimate of drug consumption would result. On the other hand, if a sample is collected when only industrial waste which is free of drugs/metabolites comes through then drug consumption in the catchment would be underestimated. As such, individual grab samples are of limited utility from the WWA perspective.

The water 'packets' can vary not just in their contents, but also in volume. For instance, the wastewater flowing *into* large WWTPs, called 'influent', can have a daily variation of 100,000 m^3 (Ort, Lawrence, Reungoat et al. 2010). That is equivalent to the volume of water contained in 40 Olympic-sized swimming pools. Clearly this variation in flow dwarfs the variation in contribution made to the wastewater stream by (the relatively few) individuals who consume drugs. Water volume is very important for WWA calculations. This is because in the laboratory, chemists quantify the mass of substances per litre of sample (i.e. the concentration of the drug), usually expressed as nanograms per litre. In a sense this can be thought of as a *ratio* of substances to water. From this analysis the total amount of substance in the 'packet' (and therefore the amount of drug consumed by the population that contributed to the wastewater 'packet') is calculated by multiplying the concentration of substances by the total volume of water in the packet. Failing to account for fluctuations in water volume flowing through the sewer at the time a sample is collected can lead to erroneous measurements. For instance, if a sample is collected when the water flow is slower than average the drug/metabolite concentration will be relatively high and an overestimate of total drug consumption will result. Conversely, failing to account for increased water volume will lead to an underestimate of drug consumption.

$$Daily\ per\ capita\ drug\ consumption_i = \frac{Concentration_i\,(in\,sample) \times Flow}{Population} \times \frac{1}{Excretion\ factor_i}$$

Ort, Lawrence, Rieckermann et al. (2010) reviewed 87 WWA publications on pharmaceuticals and personal care products. They showed that most articles provided cursory information about sampling protocols. They also contended that the literature had provided insufficient details about the protocols used to store samples and prepare them for analysis. Citing guidelines developed by the US Environmental Protection Agency in the early 1980s, Ort, Lawrence, Rieckermann et al. (2010) reminded the WWA field that inadequate handling of samples can be a cause of error. This is because chemical changes can occur to biomarkers after they are collected unless samples are preserved (see Castiglioni, Bijlsma, Covaci et al. 2016). As wastewater contains drugs and metabolites at trace concentrations it is paramount that measures to guard against contamination of samples are in place. European teams met in 2010 to form a body called Sewage Analysis CORe (SCORE). SCORE led the WWA field to rectify issues relating to sampling, analysis and reporting of results. Sara Castiglioni and colleagues (2013) collated survey data from WWTPs which had previously been involved in WWA studies across Europe. With the input of Ort and other sewage engineers, the team then drew on the survey data to develop standard protocols for sampling wastewater and handling samples prior to analysis. The 2013 guidelines are housed on the website of the European Monitoring Centre for Drugs and Drug Addiction (EMCDDA).[1] Subsequent studies have established that adopting the standard protocols reduced the margin of error attributable to sampling to 10% or less.

Autosamplers are pieces of equipment that are normally used to draw wastewater samples from the influent entering WWTPs into a refrigerated container. Typically they can be operated so that they collect samples at different intervals either in time proportional mode (where the same volume of sample is collected at fixed intervals, e.g. every 10 minutes), in flow proportional mode (where a flow signal is connected to the autosampler and the amount of volume sampled is adjusted dependent on the flow volume for a fixed period of time) or in volume proportional mode (again where the autosampler is connected to a flow signal but collects a fixed volume of sample at a fixed flow volume interval, e.g. samples every 10,000 m^3). To account for fluctuations in the *contents* of wastewater, the recommended approach is to sample in a continuous flow proportional mode – achievable with peristaltic pumps. Typically the continuous mode isn't available at WWTPs and thus the next best method is to sample flow proportionally at a high frequency, preferably at least every 10 minutes. Typically, the mini-samples – which might be 100 to 200 in number – are pooled every 24 hours and homogenised. These 24-hour composite samples may comprise many litres in volume, so a subsample of the composite is taken for analysis. One of the practical advantages of 24-hour composite samples is the fact that they represent variation over a full day, which is useful for monitoring trends over time and changes in drug consumption at weekends and so on (e.g. Lai et al. 2013).

A number of preservatives have been investigated which may be added to the samples at the point of collection to prevent changes in biomarkers that could skew results. In-sample stability and preservation technique needs to be considered on a per biomarker basis but often a compromise is required if the interest is analysing a suite of biomarkers in a sample. To prevent further degradation of biomarkers prior to analysis, samples are either frozen or extracted onto 'solid-phase extraction' cartridges, to concentrate and isolate target biomarkers from each other and other compounds. Once preserved, samples can be transported very long distances; analyses can be conducted in countries far distant from the sites where the samples were collected.

Various online resources are available to assist WWA teams to design their sampling procedures to best suit the sewerage system under study. A wide variety of factors may need to be taken into account at this stage. For example, does the sewerage system have high rates of leakage? Is it designed to carry sewage water alone, or does it also carry storm water? Do the relevant water authorities periodically transport or pump wastewater from one system to another? Do certain catchments service industrial complexes that affect influent flow? In practice, teams that are commencing WWA research internationally should seek advice from experienced scholars about these issues. The WWA field has been a collegial one to date and there are no indications of researchers being secretive or withholding their expertise.

This overview of wastewater sampling tells us several important things about the utility of WWA. First, although a high degree of expertise is needed to design an appropriate sampling process for each site, non-experts can carry out the day-to-day business of collecting and transporting samples. This is one of the reasons why WWA teams have been able to conduct studies on such a large scale to incorporate samples from dozens of cities in the same time frame.

Second, water authorities are stakeholders, even 'gate-keepers', in WWA research. Their approval is needed for sampling to take place. The authorities hold technical information about the performance of sewerage systems, including issues noted above relating to industry, storm water, pumping and so on. This information is required to optimise sampling at any given site. Additionally they also provide the flow data required for WWA calculations.

More fundamentally, water authorities have blue-prints for sewerage infrastructure: what parts of locations are serviced and what are not; the boundaries of particular catchments; and the location of WWTPs. This sewerage infrastructure is central to defining the "data collection apparatus" of WWA studies (Prichard et al. 2017, 849). Conceptualising sewerage infrastructure as an apparatus is useful. It encourages WWA teams to incorporate information about sewerage infrastructure into the development of research questions and project plans.

While water authorities can interact with WWA teams as 'stakeholders', they are better described as 'collaborative partners' in many WWA studies. Water authorities are often invested in the work of WWA studies because the research informs aspects of their core business. For example, some WWA projects also analyse wastewater *effluent* as well as influent, which assists water authorities to gauge the efficacy of their techniques to minimise adverse environmental impacts arising from the use of treated wastewater. It is not uncommon for water authorities to take carriage of sampling for WWA teams. In fact, one such collaborative relationship in Australia has successfully functioned since 2009.

2.2.3 *Laboratory Processes and Back-Calculating Population Consumption*

Thousands of hours of highly complex research has underpinned each of the advances made in WWA laboratory techniques. The disciplines involved mainly fall under the umbrella term 'analytical chemistry'. This has multiple applications in the fields of chemistry, biology, pharmacy, forensic science, environmental toxicology and engineering. A fundamental analytical chemistry technique is called chromatography. The basis of chromatography is the separation of the compounds present in complex mixtures – ranging from drugs in blood samples to hydrocarbons in crude oil to drugs in wastewater – which allows for unhindered measurement of a characteristic signal for the compounds in the mixture using some form of detector. The simplest use of chromatography is to infer the presence (or absence) of a particular compound or compounds, which is referred to as qualitative analysis. However, the size of the signal is related to the amount of compound present in the mixture and therefore (after appropriate calibration) the size of the signal can be used to measure the concentration of compounds in the mixture, which is a process called quantitative analysis.

The most common approach used in WWA is to combine a technique called liquid chromatography (LC) for separation of compounds with mass spectrometry (MS) as the means of qualitative and quantitative analysis. In effect, the signal from the mass spectrometer carries information about the size (mass) of the different molecules present in the mixture and some information about their molecular structure. As an example, if a mixture (such as wastewater) contains cocaine then a signal will be obtained that shows the presence of a mass of 304, which arises when the cocaine molecule acquires a hydrogen ion during analysis, as well as masses of 182 and 82 which arises when the cocaine molecule fragments. A signal comprising the masses 304, 182 and 82 is characteristic for cocaine.

The benefit of analytical methods like LC-MS is that they provide very high levels of sensitivity that means they are capable of detecting residues in *parts per trillion*, usually expressed as nanograms per litre. By comparison, some drug monitoring systems engage commercial laboratories to routinely analyse urine provided by arrestees for traces of illicit drugs. The drug biomarkers in unadulterated urine samples occur in concentrations that are approximately 1,000 times higher than the concentrations found in WWA samples. Other techniques are being trialled that may provide greater efficiency in the laboratory and possibly even greater sensitivity in future WWA applications.

LC-MS and related methods have proven effective despite the presence of countless types of biological and synthetic substances that enter sewerage systems (EMCDDA 2016). Not surprisingly, the LC-MS instruments are not designed to contend with faeces, rubbish and the like, so solids are removed from samples using clean-up procedures before analyses commence.

Different laboratories have different equipment and have developed different, but equally valid, approaches to analysing biomarkers in wastewater (EMCDDA 2016). The potential variability that these differences may introduce into data presented a problem when planning large-scale projects involving multiple laboratories, such as those conducted across Europe. To remedy this, inter-laboratory validation studies are conducted, similar to those conducted in other research fields, so that the data produced by participating laboratories can be calibrated for reliable comparison (Castiglioni et al. 2013; Ort et al. 2014; EMCDDA, 2018). The congruence of data produced by different laboratories has also been enhanced by the development of the SCORE protocols (see further 2.2.2), which contain guidelines for analysing wastewater samples and reporting results.

After wastewater samples are analysed for biomarkers back-calculations are used to estimate community consumption of substances. These back-calculations are similar to those proposed by Daughton (2001) (see 2.1). Castiglioni, Bijlsma, Covaci et al. (2016, 17) summarised the main steps taken in this process.

1. As noted, the laboratory tests indicate the concentrations of the biomarker(s) in the wastewater sample expressed in nanograms per litre (ng/l). This figure is multiplied by the daily flow rates of the influent entering the WWTP, which is expressed in cubic metres per day (m^3/day). The result is the 'daily load' – grams per day (g/day) – of substances in the wastewater stream.
2. Total consumption is estimated using a correction factor, which is based on information about the average metabolic processing of each substance, including what biomarkers are excreted, the ratio of biomarkers to each other, and the average rate of excretion.
3. The daily consumption estimate is divided by the estimated catchment population size (the number of people serviced by the WWTP). This results in an estimate of milligrams per day per 1,000 people.

Many WWA publications present data based on this level of calculation – usually presenting these estimates over multiple days to show temporal trends.

4. For any given substance a fourth step may be to estimate how many doses were consumed. This involves dividing the estimated milligrams consumed (per day per 1,000 people) by the 'standard' dose in milligrams. For example, if WWA shows that consumption of a particular drug was 200 mg/day/1,000 people and the standard dose of that drug is 100 mg, then the estimate is that two doses of the drug were consumed per day per 1,000 people.

It is clear that WWA data measure *consumption rates* in large groups of people; they do not inform us about the use of drugs by *individuals*. This is the cardinal limitation of the WWA method that is explored further below in 2.4. It is also ironically WWA's main strength with respect to research ethics as discussed in 2.3.

2.2.4 *Factors Relevant to Interpreting WWA Data*

The WWA field is wary about communicating results without including caveats about the interpretation of its metrics. Perhaps this is partly due to the scale and nature of the media coverage of the first paper that reported back-calculation estimates of cocaine use (Zuccato et al. 2005, see 2.1)? The European Monitoring Centre for Drugs and Drug Addiction produced detailed reports in 2008 and in 2016, both of which were co-authored by key members of SCORE (see EMCDDA 2016). These reports catalogue key factors that are relevant in interpreting WWA data. It is important to review these factors to understand the utility of WWA data and its current limitations. Given the intensity of research that is focused upon overcoming these limitations, we agree with Mounteney et al. (2015) that WWA is likely to become a more important source of information on illicit drug use in the future.

2.2.4.1 Population Size

It can be difficult to estimate population size in a WWTP catchment area. If researchers overestimate population size they will underestimate the amount of substance consumed per day per 1,000 people in the catchment. Underestimating population will have the opposite result.

Key terms used by statisticians on this topic are *de facto* and *de jure* populations. The first term refers to the actual number of people in an area at any given time. The second term refers to the number of people who reside in the area. *De jure* figures may be adequate for many national and local policy objectives. However, *de jure* figures are not ideal for WWA particularly because they do not account for *population movement* between catchments. Obviously populations can fluctuate very significantly in urban settings as people travel to work during the day. Nightclub districts, for example, are especially likely to have large changes in population size in the course of an evening.

Various techniques are being trialled to better approximate the size of populations (e.g. Been et al. 2014; Castiglioni et al. 2013; Gao et al. 2016; O'Brien et al. 2014; Thomas et al. 2012). The hypothesis driving these trials is that there could be a substance (or substances) present in wastewater that everyone (or at least an accurately known fraction of the population) deposits into the wastewater at an accurately known daily rate. If such substances could be quantified daily in wastewater then a reliable surrogate for population size would be achieved. Lai et al. (2011) demonstrated that it was possible to estimate population by focusing on levels of pharmaceuticals and commonly ingested substances such as artificial sweeteners in wastewater samples. The approach involved first estimating the consumption of these substances using WWA and triangulating the results with other sources of data, including prescriptions of pharmaceuticals and sales data for artificial sweeteners. Other studies have examined the utility of biomarkers that are regularly produced by humans as a proxy for population size (e.g. Thai et al. 2014b). These methods effectively use the information contained in the wastewater to calculate population size (see further Gracia-Lor et al. 2017). In Australia, the national census data can collect both *de jure* and *de facto* data on a specific date. Researchers are examining how this information may be used to develop better models for accurately estimating *de facto* population size for WWA data using wastewater samples themselves (see O'Brien et al. 2014; Lai et al. 2015). An alternative approach has also been trialled where population changes were monitored using mobile phone tower ping data in Oslo (Baz-Lomba et al. 2019).

2.2.4.2 Purity, Potency and Standard Doses

For many substances widely accepted (and static) metrics exist as to what constitutes a 'standard dose'. Examples include medications and legal products like alcohol. By contrast, a dose of illicit drugs can fluctuate for a wide variety of reasons (Brunt et al. 2015). The illicit market usually reacts to fluctuations in purity by changing prices. For instance, downward fluctuations can lead to lower prices and smaller standard doses; the reverse can occur if drug purity increases. Decreases or increases in purity can lead to consumers receiving respectively less or more of an illicit drug – at least for a period of time. These trends may be detected in WWA. It is therefore important for purity to be considered when interpreting WWA data (Bruno et al. 2014). Researchers can access relatively good information about standard doses in any given year. Forensic laboratories that support police agencies may be able to provide WWA researchers with up-to-date intelligence about purity. In Europe very accurate information can be accessed from the Trans-European Drug Information project, which analyses tens of thousands of illicit drug doses provided by drug users (Brunt et al. 2015).

While users have little control over what the market provides, users do have control over how much they consume per dose. It is known that experienced users can tolerate higher quantities of drugs than inexperienced users and therefore what is a single dose to one user may be multiple doses to another and it follows that each will contribute different amounts to the wastewater stream after a single dose.

Illicit drugs also differ in their *potency*. Potent drugs require a smaller dose to affect the human body than do less potent drugs. For example, in 2009 in Australia the standard dose of methamphetamine was 41 mg yet the standard dose for cocaine was 140 mg (Prichard et al. 2012). Load data conceal these differences because they only present the estimated mass consumed per day per 1,000 people. Estimating doses consumed per day per 1,000 people takes potency into account and facilitates comparisons of the relative strengths of drugs in different drug markets.

This was demonstrated in an early Australian study (Prichard et al. 2012). Analyses of samples collected in 2009 indicated that 158 mg of methamphetamine and 221 mg of cocaine were consumed daily per 1,000 people. However, this translated to 3.9 doses of methamphetamine and 1.6 doses of cocaine – indicating that methamphetamine was the stronger of the two markets during the time of sampling.

Dose-estimation can be used to assist in understanding the relative economic value of different illicit drug markets. This simply involves multiplying the number of doses estimated to be consumed daily per 1,000 people by information about the 'street value' of doses. Prices fluctuate, but good annual figures can be sourced from drug monitoring systems in different regions.

2.2.4.3 Correction Factors

Correction factors require data on excretion rates for all substances, including illicit substances. The data concerning the excretion of illicit substances are derived from a relatively small body of work – the accuracy of which is questionable (Castiglioni, Bijlsma, Covaci et al. 2016). Large-scale dose-elimination experiments involving illicit substances are difficult to carry out and therefore results from such trials are rare in the literature. Other data originate from forensic toxicological examination of fatal drug-related incidents, but these cases often involve polydrug use or acute overdose, neither of which may be applicable to excretion of drugs after chronic administration of non-fatal, recreational doses. In any event, the excretion data available in the literature are average figures derived from relatively small survey populations. The survey population size is relevant because the rate and extent of metabolisation varies from individual to individual and the best way of ensuring that the data obtained are not perturbed by inclusion of date from extreme individuals is to ensure that the population data represent 'normal' individuals, which means testing a large population. While there are implications with regards to using average data from small populations when estimating drug consumption in a large metropolitan community, the situation becomes more important when estimating consumption in a small community, such as prisons or small towns. This is because in a small community there may be a user community of only a few or few tens of individuals. One or two extreme individuals in that population will have a big influence upon the community estimate.

As noted in 2.2.1, the *products* excreted by people after they use illicit drugs do not differ by route of administration (smoking, injecting, swallowing, etc.) but the *rate* of excretion can vary by route (Prinzleve et al. 2004). Excretion rates can also differ depending on the frequency at which individuals use illicit drugs.

2.2.4.4 Complexities Associated With Some Illicit Drugs

Among the major illicit drugs, **cocaine** has proven to be the simplest for WWA scientists to monitor (Khan and Nicell, 2011). Its biomarkers are the parent compound (cocaine) and a metabolite, benzoylecgonine, which is uniquely produced by the human body when excreting cocaine (see further 2.2.1). There is good confidence also in the WWA field about the standard approaches for monitoring **MDMA** consumption (see Kasprzyk-Hordern and Baker 2012). Similarly, few problems are reported in measuring **LSD** or **ketamine.**

The illicit drug that has presented difficulties for WWA is **heroin.** Back-calculations of heroin consumption using O^6-acetylmorphine as the biomarker may not be fully quantitative due to in-sample and in-sewer stability issues (McCall et al. 2015; Thai et al. 2014b) and thus present difficulties with spatial comparisons. However the addition of sodium metabisulfite to samples has been identified to increase the in-sample stability of O^6-acetylmorphine in wastewater (Tscharke et al. 2016) and, thus, assuming catchment hydraulic residence times remain consistent, monitoring temporal trends in heroin consumption using WWA may still be achievable. For instance, the National Wastewater Drug Monitoring Program in Australia includes O^6-acetylmorphine monitoring (ACIC 2019).

The parent compound is rarely found in urine because the drug is metabolised quickly. The main metabolite of heroin is O^6-acetylmorphine, which in turn is rapidly converted into morphine. Unfortunately morphine is excreted after consumption of the important opiates morphine and codeine (which up until recently was sold without a prescription in Australia). Back-calculation might be feasible if sales or prescription data were available on codeine consumption. Other problems relate to the fact that morphine is not stable in sewage water (EMCDDA 2016) and that O^6-acetylmorphine is very hard to detect in wastewater due to its instability in water.

Several different issues affect the reliability of analyses of levels of **cannabis** with WWA. The main metabolite of the active ingredient of cannabis, tetrahydrocannabinol (THC), is THC-acid. This compound is excreted at relatively low levels (Castiglioni, Bijlsma, Covaci et al. 2016). It is also lipophilic which means that it binds to solid fats contained in faeces and so is underrepresented in wastewater (Khan and Nicell 2011). These dynamics are taken into account in the correction factor for THC-acid, but it means that there are greater uncertainties in estimating levels of cannabis use from wastewater than is the case for other licit and illicit drugs.

Wastewater studies regularly monitor both **methamphetamine** and **amphetamine**. However, the back-calculation process needs to take into account each country's consumption of pharmaceuticals as well as patterns of illicit drug use. A challenge for WWA researchers is that methamphetamine is partly excreted unchanged and partly metabolised into amphetamine (Baselt 2008). In addition, both methamphetamine and amphetamine are metabolic by-products of some pharmaceuticals (EMCDDA 2016), such as those used to treat attention deficit hyperactivity disorder (dexamphetamine) and Parkinson's disease (selegiline) (Mastroianni et al. 2017). The black market in Australia, like the USA and parts of Asia, is dominated by methamphetamine (ACIC 2019). In those jurisdictions where forensic laboratories have shown that the consumption of amphetamine is negligible, there is no need for it to be corrected-for in back-calculations of methamphetamine. In Australia, when back-calculating methamphetamine researchers have subtracted the estimated contribution of dexamphetamine and selegine using national prescription data (Prichard et al. 2012). Interestingly, this step is not necessary in Europe because dexamphetamine and selegiline are not widely prescribed there (Mastroianni et al. 2017).

New psychoactive substances (NPS) include hundreds of substances that mimic the physiological effects of major classes of illicit drugs but fall outside of the definitions used by the criminal law in different countries. Some NPS, such as mephedrone and pentedrone, have established appreciable markets in North America and Europe (Thai et al. 2016). Major obstacles that have hampered the use of WWA to monitor NPS consumption include: the lack of information regarding the metabolisation of NPS (rate of excretion and potential biomarkers); the comparatively small scale of the NPS market and hence low levels of specific NPS in wastewater; and the extensive diversity of NPS (Reid and Thomas 2016).

2.2.4.5 Dumping Illicit Drugs in Sewers

One of the most common questions asked about WWA by audiences is whether the results can be biased if illicit drugs are flushed down toilets by people wanting to avoid detection by police. Standard WWA cannot distinguish between methamphetamine, for example, that has been flushed down sewers and methamphetamine that has passed unchanged through the body after its administration. One strategy that is used to address this issue is to examine the correlation between the parent compounds and the metabolites that have been chosen as biomarkers. Significant increases in parent compound loads (e.g. cocaine or methamphetamine) that are not matched by metabolite loads (i.e. benzoylecgonine or amphetamine, respectively) may indicate that direct disposal to the sewer (dumping) has occurred.

A recent enhancement of WWA, enantiomeric profiling, may also be used to differentiate between consumption and dumping of those drugs that exist in enantiomeric forms (also referred to as optically active forms), which applies to the important substances amphetamine, methamphetamine and MDMA (Emke et al., 2014; Petrie et al., 2016). For these drugs their molecular structures are such that they can exist in two forms that are mirror images of each other but are not identical. A simple analogy involves letters of the alphabet. The mirror images of letters such as 'O', 'A' and 'V' are identical to the original letters whereas the mirror images of letters such as 'R' and 'Q' are quite different to the original letters. Another example involves shoes; a right shoe is not identical to a left shoe, even though they are constructed in very similar ways. A right shoe and a left shoe are non-identical mirror images of each other and parts of our body that are also related as non-identical mirror images (our feet) are able to sense the difference between shoes. A drug-related example is the infamous substance thalidomide. It can occur in two molecular forms that are mirror images of each other. One molecule was quite effective in treating the symptoms of morning sickness and was otherwise benign – unfortunately the mirror image molecule that was present in the dose was far from benign and caused major birth defects. The application of enantiomeric profiling to WWA analysis is based on the fact that illicit manufacture of some important drugs (notably amphetamine and MDMA) will usually produce the two mirror image molecules in equal amounts (which is called a racemic mixture). However, when such a mixture is consumed the human body will process the two differently, resulting in the depletion of one of the forms in the excreted drug mixture. On the other hand, if amphetamine or MDMA are dumped into wastewater the two forms will be present in identical concentrations and this perturbation of the usual signature could be used by authorities to identify situations where wastewater data are likely to be inflated by dumping. Such an approach is complex, however. One complexity arises from the assumption that illicit material is a mixture of the two enantiomeric forms in equal amounts. This assumption is not valid in many jurisdictions where ephedrine or pseudoephedrine are used as precursors for methamphetamine, as only one enantiomer of the precursor is used and one enantiomer of methamphetamine is produced by that process. In those jurisdictions where illicit drug manufacturers deliberately enrich their product in one enantiomer the same complexity will arise. In regard to amphetamine measurement, the approach can be slightly more complicated if there is a significant community usage of prescribed amphetamine or pharmaceuticals that are converted into it by the body. This is because some of these preparations have only one of the two enantiomers present and some have both. The usage of prescription amphetamine perturbs the ratio of the two amphetamine molecules in wastewater, which could obscure a dumping. In any event for any drug, enantiomeric profiling would be effective only when dumping involves relatively large quantities of material (Prichard et al. 2012). Patently, the value of illicit drugs to sellers means that it is unlikely that there will be regular dumping of quantities sufficient to affect WWA data.

Nevertheless, dumping does occur as illustrated by a spectacular example discussed by Castiglioni, Bijlsma, Covaci et al. (2016). They found a 20-fold increase in MDMA between 2010 and 2011 in the city of Utrecht in The Netherlands (Bijlsma et al. 2012). According to Castiglioni, Bijlsma, Covaci et al. (2016), the likely cause of the increase was the dumping of 30 kg of MDMA by drug traffickers during a raid by law enforcement officers. Since the average MDMA dose in The Netherlands in 2010 was about 100 mg (Brunt, Cross and Peck 2015), 30 kg equates to 300,000 doses. In today's terms in street sales this would be worth approximately €6 million,[2] or USD7.34 million. Interestingly the peak took several days to abate. But this seems to be attributable to the systems used at the WWTP in Utrecht. It 're-treated' some effluent by pumping it back into the influent stream (Castiglioni, Bijlsma, Covaci et al. 2016).

The Utrecht case study underscores the value of collaborative relationships between WWA teams and police and water authorities in interpreting WWA data – especially when anomalies are encountered. As a final point, it is heartening for the WWA field that law enforcement agencies – that might reasonably be expected to have expert knowledge of the frequency and scale of dumping – are becoming interested in WWA data. The most notable example is the Australian Criminal Intelligence Commission (ACIC), which has funded WWA monitoring across Australia since 2016. This commitment demonstrates that the ACIC does not consider that the efficacy of WWA is substantively weakened by the issue of dumping or any of the limitations discussed above.

2.3 Human Research Ethics

As the field matured it became apparent that WWA did not neatly fit under existing national codes or guidelines for research. Despite the fact that WWA studies human waste, it is impossible to construe the data that WWA gathers as data on individuals (Hall et al. 2012). An individual's urine is intermingled with urine from many thousands of people, not to mention gigalitres of water and a multitude of synthetic compounds. This context does not raise the same ethical concerns as raised by more traditional research on drug use within criminology, epidemiology, public health surveillance or environmental health research.

Assessing the ethical dimension of any area of human research requires consideration of four main principles, summarised in Table 2.1.

Hall and colleagues (2012) have argued that WWA satisfies these ethical principles when it is used to study the general population. Although individuals' data are analysed without consent, the impossibility of identification means that the principle of autonomy is not threatened. Since the objective of WWA is to inform practices that aim to reduce the harms associated with drug use it can be said to meet the principle of beneficence. Distributive justice is served inasmuch as burdens and benefits are

Table 2.1 Principles of Ethical Research

Respect for participants' personal autonomy	Typically demonstrated by requiring that: individuals must provide fully informed consent before they can participate in research; and that participants' confidentiality and privacy is maintained	Beauchamp and Childress, (2001); Fry and Hall (2004)
Beneficence	Requires that research has a reasonable prospect of producing benefits to the participants, or at least to society in general	Beauchamp and Childress (2001)
Distributive justice	Requires a fair and equitable distribution of the burdens and benefits of research participation	Brody (1998)
Non-maleficence	Requires researchers to avoid placing participants at the risk of harm or to minimise harm caused	Brody (1998)

Source: Adapted from Prichard et al. 2014.

shared across entire catchments. Regarding non-maleficence, it should be noted that the concept of 'harm' here has a considerably lower threshold than at civil or criminal law. For example, in the research context harm can be taken to include 'distress', 'embarrassment' or 'stigmatisation' (e.g. NHMRC, ARC and Universities Australia 2007, 16). However, even under such definitions the risks of harms are usually very low.

Only one ethics committee in the world has required a review of a WWA study which it approved as low-risk. The WWA field nonetheless recognised that the lack of ethical oversight by external bodies necessitated that the field develop ethical research guidelines (Prichard et al. 2014). These were developed by 15 scholars across Europe and Australia (Prichard et al. 2016). They are housed on several websites, including the EMCDDA's (see SCORE 2016).

For general population studies, the guidelines concentrated on potential harms that sensationalist media coverage of WWA findings might have upon particular groups – as identified by Hall et al. (2012) and Prichard et al. (2010; 2014). These risks include: shame or embarrassment for vulnerable communities (e.g. in suburbs or towns with high levels of social disadvantage); amplifying the stigmatisation of vulnerable groups; encouraging populist responses to the drug-crime nexus that may contribute unnecessarily to 'net-widening' (Austin and Krisberg 1981, 165) (e.g. by bringing drug users into the criminal justice system); and causing economic harm by adversely affecting the reputation of communities or events, such as festivals (see Prichard et al. 2016, 8). Some of the nuances relating to small communities and towns, particularly in the rural context, are explored in more detail in Chapter 4.

The guidelines recommended that WWA teams develop media-communication strategies to ensure that data were not misreported. They also proposed that researchers consider anonymising sites involved in their studies. If necessary this should involve removing any information from publications that might lead to the identification of sites, such as population size and general location. Finally, it is worth noting that the guidelines underscore that WWA teams must ensure that they (a) comply with the laws and regulations of their countries and institutions and (b) have the consent of water authorities before sampling from their facilities.

Ethical complexities increase markedly where WWA is used to study particular buildings, such as prisons, schools and workplaces. These are discussed in Chapter 5.

2.4 *Applications of Wastewater Analysis*

It is not the purpose of this section of the chapter to exhaustively catalogue WWA studies according to their objectives and findings (see generally

Castiglioni and Vandam 2016). This section overviews certain highpoints in the wastewater field that illustrate the scale at which the method can be applied. It acts as a forerunner to Section 2.5 which is a compendium of the strengths and shortcomings of WWA comparative to other drug research methods. Subsequent chapters will examine in more detail WWA research conducted at the international (Chapter 3), national (Chapter 4) and local (Chapter 5) levels.

WWA studies have been published at a steady rate and WWA has been applied in an increasing number of countries. Europe, Australia and North America aside, WWA studies have been conducted in Taiwan (Lin et al. 2010), China (Khan et al. 2014; Lai et al. 2013), South Africa (Archer et al. 2018), Colombia, the West Indies and elsewhere (see further Chapter 3). As already explained, the major focus of WWA is monitoring illicit drug consumption in the general population with the objective of supplementing, not supplanting, traditional methods.

Large-scale WWA drug monitoring studies were first trialled in Oregon (Banta-Green et al. 2009) and Belgium (van Nuijs et al. 2009). More recent national studies have included Finland and Sweden (EMCDDA 2016) and Australia (Lai et al. 2016) – the latter project analysing samples from catchments servicing approximately 40% of the Australian population.

The largest scale application to date has been the European WWA monitoring system. This is conducted by SCORE under the aegis of the EMCDDA. SCORE has conducted international synchronised studies annually since 2011 (Thomas et al. 2012). This monitoring system is an impressive logistical feat made possible in large part by the protocols SCORE developed for data collection and analysis, and because SCORE conducts repeated inter-laboratory calibration tests (see 2.2.3). The first phase in 2011 involved 19 cities in 11 countries. The number of sites included in each phase has grown each year. The 2017 phase sampled from 58 cities in 19 countries, including the Czech Republic, Slovakia and Cyprus (EMCDDA 2018).

The results from the SCORE program show a good level of agreement with data from the other European drug monitoring systems in terms of the: rank order of the consumption of drug types (cocaine, amphetamine, MDMA, methamphetamine and cannabis); changes over time; and variance within and between countries (Castiglioni, Vandam and Griffiths 2016; Ort et al. 2014). The general concordance between WWA and other drug monitoring methods has been reported elsewhere. It is widely interpreted as evidence that the trends observed in WWA data accurately reflect true population levels of drug consumption (EMCDDA 2018).

The EMCDDA (2018) is confident that WWA will prove to be advantageous in predicting changes in population-level drug consumption. This confidence seems reasonable given that, compared with other monitoring methods, WWA can produce data quickly, at a higher frequency and on a large scale (Ort et al. 2014). It can also be used to generate data where none previously existed, particularly in regional towns (e.g. Prichard et al. 2018). Perhaps the best example where WWA has been the first method to detect a change in drug consumption is a study by Zuccato et al. (2011), which accurately reported a downturn in consumption of cocaine and heroin in two Italian cities before the trend appeared in other monitoring systems (Castiglioni and Vandam 2016).

In addition to monitoring, WWA has examined the relationship between community drug consumption patterns and times, seasons and events. The spike in drug consumption on weekends – particularly for party drugs like MDMA – has been well documented in WWA studies (Ort et al. 2014; Prichard et al. 2012). Other publications have investigated patterns of drug consumption during holiday periods (Lai et al. 2013) and televised sporting events (Gerrity et al. 2011).

In recent years greater attention has been paid to estimating the population consumption of alcohol (Boogaerts et al. 2016; Mastroianni et al. 2017; Reid et al. 2011; Rodríguez-Álvarez et al. 2015) and tobacco (Castiglioni et al. 2015; Lopes et al. 2014; Wang and Chu 2016). The target biomarker for alcohol is ethyl sulfate and for nicotine, cotinine and trans-3′-hydroxycotinine (Lai et al. 2018). The processes for analysing samples and back-calculating consumption rates are the same as described earlier in 2.2.3 for illicit drugs. The capacity of WWA to provide spatio-temporal mapping of alcohol and tobacco consumption over large areas has been established; good levels of agreement have been found between WWA figures and representative surveys of the general population (Lai et al. 2018). Later chapters will examine the implications of these newer applications of WWA for measuring the impact of policies at all levels (macro, meso and micro), including prisons.

General population studies aside, WWA is used periodically to examine drug use on a very small geographical scale. Most site-specific studies sample wastewater from particular buildings, such as prisons (van Dyken et al. 2014) and schools (Zuccato et al. 2017). Others have examined samples from the small sewerage treatment systems established for music festivals (Lai et al. 2013).

2.4.1 Options for New Researchers

The WWA literature shows that laboratories invest substantial amounts of time refining techniques to conduct WWA analyses according to best practice standards (see further 2.2.3). The equipment used to undertake these tests are very expensive and consequently are often in high demand by multiple teams with different research objectives. Preparing and conducting WWA obviously also requires technical knowledge of sampling procedures, relationships with water authorities and an understanding of relevant research questions related to illicit drug use. The latter requires interdisciplinary collaboration. The most productive WWA teams have benefitted from partnerships between several disciplines. For these and many other reasons, research teams are unlikely to form with the intention of applying WWA in a single or short-term project. WWA requires tangible and intangible infrastructure that is not easily assembled.

Non-scientists who want to move into the WWA field have two main options. They can initiate discussions with established WWA teams – keeping in mind that it is feasible for samples to be transported many thousands of kilometres for laboratory analysis. Or they can investigate whether other laboratories are interested in building capacity to undertake WWA analyses. Scientists working in analogous areas of analytical chemistry may be able to transition to WWA. They will find the guidelines and protocols on data collection and analysis referred to earlier very useful. They may also choose to make contact with experienced WWA teams to obtain assistance and advice.

2.5 Drug Research Methods: Comparative Strengths of Traditional Approaches and Wastewater Analysis

Chapter 1 reviewed the key methods that are used internationally to monitor the prevalence of illicit drug consumption. The task is very difficult because drug use is illegal in most jurisdictions and activities relating to supply and demand in the market remain hidden by choice. For this reason almost all reliable sources of information about drug consumption are drawn upon and triangulated to inform agencies about the performance of their policies and practices. It is hard to locate any WWA publications which suggest that WWA can replace existing methods. On the contrary, the message the field has uniformly projected is that WWA has strengths complementary to existing methods and consequently is a useful additional tool for drug monitoring.

Table 2.2, provides a schedule of the main strengths and limitations of measuring drug consumption with established methods and WWA. It is adapted from a table developed by Castiglioni et al. (2014, 618). The picture projected by Table 2.2 is that the strengths of established methods compensate for deficiencies in WWA and vice versa.

Regarding *Item 1* of Table 2.2, scholars have asserted or implied that WWA is inexpensive in comparison to established survey methods (e.g. Prichard et al. 2012; Castiglioni et al. 2014; Castiglioni, Zuccato, Thomas and Reid 2016). However, clear metrics on the topic are not readily

Table 2.2 Relative Strengths of Established Methods and WWA for Monitoring Illicit Drug Consumption

Issue #		Established Methods	Wastewater Analysis
1	High costs of studies	Yes	No
2	Rapid estimates	No	Yes
3	Objective estimates	Varying	Yes
4	Maximum sample sizes	Tens of thousands	Millions
5	Regional research	Limited	Yes
6	Retrospective analysis	Limited	Yes
7	Ethical protocols	Complex	Simple
8	Emergence of new drugs	Potential	Potential
9	Individuals' patterns of substance use	Yes	No
10	Route of administration	Yes	No
11	Drug-related harms	Yes	No
12	Purity of drugs	Yes	No

Source: Table modified from Castiglioni et al. 2014, used with permission.

available. This is partly because WWA has not been fully costed (e.g. to take account of in-kind assistance or the labour of research students) and partly because agencies that operate many of the established methods of drug monitoring do not publicly report the expenses involved in using these methods. The resources needed to conduct survey studies vary. For example, one-on-one interviews (either in person or by telephone) require considerably more resources than online surveys. Similarly, the effort required to conduct statistical analyses of law enforcement data depends heavily upon the steps required to prepare the data for analysis (also called data cleaning).

Approximate information on the annual Australian monitoring systems indicates that the country's general population study – the National Drug Strategy Household Survey – costs AUD2.4 million (€1.5 million, USD1.9 million) but is only conducted every three years (AusTender 2013). A combination of surveys of specific groups of drug users and analyses of secondary health and law enforcement data, undertaken by the National Drug and Alcohol Research Centre, annually costs approximately AUD1.4 million (€0.9 million, USD1.1 million) (NDARC 2017, 64).[3] The ACIC funded three years of national WWA monitoring (2016–19) with AUD3.6 million – an average annual cost of AUD1.2 million (€0.8 million, USD0.9 million) (ACIC 2018). These broad figures generally support the claim that WWA is cost-effective. Yet, future research that precisely models economic costs of different systems would be beneficial for policy makers and funders.

Items 2–8 of Table 2.2 are not contentious on the whole. Large survey studies – on illicit drugs, alcohol and tobacco – that are representative of the general population (as opposed to convenience samples) can reach tens of thousands of people over a period of years. However, they are subject to sampling selection effects with response rates typically below 50%. Error can also affect survey data because of inaccuracies in the subjective information provided by participants. It can be extremely difficult to recruit *representative* samples for surveys of people in regional areas (Prichard et al. 2018).

Statistics on drug-related arrests and seizures are reputable indicators of *actual* drug market activity (e.g. Prichard et al. 2018). However, law enforcement data are burdened by duality; it is hard to disentangle the extent to which changes in rates of arrest or seizures represent changes in drug consumption or increased activity by law enforcement agencies (Willis, Anderson and Homel 2011).

By comparison to household surveys that are conducted every few years, WWA can provide raw data to authorities about two months after sampling from catchments that service millions of people. This means WWA can give robust evidence about trends and the rank order of substance types according to their consumption prevalence (ACIC 2019). Notwithstanding the uncertainties involved in WWA methods (detailed above in 2.2), WWA data are more objective than and less affected by the selection effects relevant to surveys. WWA in many regional communities can be conducted as easily as in urban settings, providing the community is serviced by a WWTP (Prichard et al. 2018).

Laboratories that commit to generating a frozen archive of wastewater samples over time can conduct retrospective WWA analyses. Only a fraction of each sample is needed for laboratory tests and the rest can be labelled and stored for re-analysis in the event that the emergence of a new drug is detected by other means such as surveys or coronial and clinical toxicological examinations. Current systems can only monitor new substances if they are consumed in sufficient quantities within a catchment and if information is available about how the substances are metabolised (see 2.2.4). The WWA field is investigating how to improve sensitivity of analysis to permit WWA to detect new drugs at low levels of use in the population (Reid and Thomas 2016)

Researchers using established methods of drug monitoring routinely follow well-designed protocols that, among other things, respect participants' autonomy and manage the risks of harming participants. However, the development and application of these protocols is often complex. Section 2.3 explained why WWA to monitor the general population is comparatively simple – mainly because individuals cannot be identified.

Items 9–12 in Table 2.2 summarise the major limiting characteristics of WWA. In market terms, WWA measures macro behaviour but reveals nothing about individual 'consumers' or their number. It cannot reveal whether a catchment has 1,000 heavy users, or 10,000 light recreational users. Polydrug use (including use of alcohol and legal use pharmaceuticals) is largely impossible for WWA to study meaningfully. Routes of administration cannot be discerned by WWA, at least with current techniques, and drug purity cannot be estimated using the method. Perhaps most critically for criminologists and epidemiologists, WWA omits the narratives of individuals' substance use: their frequency of use; whether they use during pregnancy; how it affects their health; its impact on relationships, education, employment and housing; its interrelationship with criminal behaviours and so forth.

2.6 Conclusion

Elsewhere we have explained why WWA represents a paradigm shift in conceptualising drug monitoring (Prichard et al. 2017). Until now science has played an important supportive role in drug monitoring systems and this was touched upon in Chapter 1. In particular, existing systems collate various sorts of data from scientific laboratories, such as the identity of illicit drugs seized by police, the types of drugs found in arrestees' urine, and the presence of alcohol and drugs in the bodies of road accident victims. With WWA, data are derived by analytical chemistry – science is the central explanatory tool. WWA data are novel and probably foreign to the social scientist. Other features of WWA require new ways of thinking too, like viewing sewerage infrastructure as a data collection tool and recognising water authorities as research partners.

This chapter has attempted to tackle the novelty of WWA by providing non-scientists with information that may allow them to understand its core features. Further it has been argued that the full potential of WWA will only be reached with greater engagement between researchers and policy makers from different backgrounds. This view is shared with others in the WWA field (Castiglioni, Vandam and Griffiths 2016).

Some literature has called for investigations into new ways to utilise WWA. Could WWA be used in intervention studies to measure the impact of supply or demand reduction strategies (Castiglioni, Vandam and Griffith 2016) – particularly in rural communities (Prichard et al. 2018) and prisons (Prichard et al. 2010)? Can WWA data be analysed in tandem with other data sources to address complex research or policy questions (e.g. Lai et al. 2018)? The following chapters respond to this call by systematically considering the potential application of WWA in a wide variety of policy relevant contexts.

Notes

1. See www.emcdda.europa.eu/activities/wastewater-analysis, Key Documents, *SCORE consensus protocol for sampling, analysis and reporting*.
2. Based on an average cost of €20 per dose in The Netherlands (van der Gouwe et al. 2017).
3. This estimate based on a reported cost of AUD6,916,797 between 2012 and 201 7.

References

Ahel, M., and W. Giger. 1985. Determination of alkylphenols and alkylphenol mono-and diethoxylates in environmental samples by high-performance liquid chromatography. *Analytical Chemistry* 57:1577–83.

Archer, E., E. Castriganò, B. Kasprzyk-Hordern, and G. M. Wolfaardt. 2018. Wastewater-based epidemiology and enantiomeric profiling for drugs of abuse in South African wastewater. *Science of The Total Environment* 625:792–800

AusTender. 2013. Contract Notice View – CN153521: 2013 National Drug Strategy Household Survey. www.tenders.gov.au/?event=public.cn.view&CNUUID=98FE2178-FB1B-23E2-2B550BC041C1C99C.

Austin, J., and B. Krisberg. 1981. NCCD research review: Wider, stronger, and different nets: The dialectics of criminal justice reform. *Journal of Research in Crime and Delinquency* 18:165–96.

Australian Broadcasting Corporation. 2005. Cocaine courses through Italian waterway. www.abc.net.au/news/2005-08-06/cocaine-courses-through-italian-waterway/2074750.

Australian Criminal Intelligence Commission. 2018. National Wastewater Drug Monitoring Program – Report 5. www.acic.gov.au/sites/default/files/nwdmp5.pdf?v=1564718845

Australian Criminal Intelligence Commission. 2019. National Wastewater Drug Monitoring Program – Report 7. www.acic.gov.au/sites/default/files/2019/06/nwdmp7_140619.pdf?v=1560498324.

Banta-Green, C. J., A. J. Brewer, C. Ort, D. R. Helsel, J. R. Williams, and J. A. Field. 2016. Using wastewater-based epidemiology to estimate drug consumption – Statistical analyses and data presentation. *Science of the Total Environment* 568:856–63.

Banta-Green, C. J., and J. A. Field. 2011. City-wide drug testing using municipal wastewater: A new tool for drug epidemiology. *Significance* 8:70–74.

Banta-Green, C. J., J. A. Field, A. C. Chiaia, D. L. Sudakin, L. Power, and L. De Montigny. 2009. The spatial epidemiology of cocaine, methamphetamine and 3,4-methylenedioxymethamphetamine (MDMA) use: A demonstration using a population measure of community drug load derived from municipal wastewater. *Addiction* 104:1874–80.

Baselt, R. C. 2008. *Disposition of toxic drugs and chemicals in man*. 8th ed. Foster City: Biomedical Publications.

Baz-Lomba, J. A., F. Di Ruscio, A. Amador, M. Reid, and K. V. Thomas. 2019. Assessing alternative population size proxies in a wastewater catchment area using mobile device data. *Environmental Science & Technology* 53 (4):1994–2001.

BBC News. 2005. Italian river 'full of cocaine'. http://news.bbc.co.uk/2/hi/europe/4746787.stm.
Beauchamp, T. L., and J. F. Childress. 2001. *Principles of biomedical ethics* (5th ed.). New York: Oxford University Press.
Been, F., L. Rossi, C. Ort, S. Rudaz, O. Olivier Delémont, and P. Esseiva. 2014. Population normalization with ammonium in wastewater-based epidemiology: application to illicit drug monitoring. *Environmental Science and Technology* 48:8162–69.
Bhattacharya, S. 2005. River flowing with cocaine indicates 'vast' drug use. Accessed 6 March 2018. www.newscientist.com/article/dn7798-river-flowing-with-cocaine-indicates-vast-drug-use/.
Bijlsma, L., A. M. Botero-Coy, R. J. Rincón, G. A. Peñuelac, and F. Hernándeza. 2016. Estimation of illicit drug use in the main cities of Colombia by means of urban wastewater analysis. *Science of The Total Environment* 565:984–93.
Bijlsma, L., E. Emke, F. Hernández, and P. de Voogt. 2012. Investigation of drugs of abuse and relevant metabolites in Dutch sewage water by liquid chromatography coupled to high resolution mass spectrometry. *Chemosphere* 89(11): 1399–1406.
Boogaerts, T., A. Covaci, J. Kinyua, H. Neels, and A. L. N. van Nuijs. 2016. Spatial and temporal trends in alcohol consumption in Belgian cities: A wastewater-based approach. *Drug and Alcohol Dependence* 160:170–76.
Brewer, A. J., C. J. Banta-Green, C. Ort, A. E. Robel, and J. Field. 2016. Wastewater testing compared with random urinalyses for the surveillance of illicit drug use in prisons. *Drug and Alcohol Review* 35:133–37.
Brody, B. A. 1998. *The ethics of biomedical research: An international perspective.* New York: Oxford University Press.
Bruno, R., W. Hall, K. P. Kirkbride et al. 2014. Commentary on Ort et al. (2014): What next to deliver on the promise of large scale sewage-based drug epidemiology? *Addiction* 109:1353–54.
Brunt, J., K. L. Cross, and M. W. Peck. 2015. Apertures in the *Clostridium sporogenes* spore coat and exosporium align to facilitate emergence of the vegetative cell. *Food Microbiology* 51:45–50.
Castiglioni, S., L. Bijlsma, A. Covaci et al. 2013. Evaluation of uncertainties associated with the determination of community drug use through the measurement of sewage drug biomarkers. *Environmental Science and Technology* 47:1452–60.
Castiglioni, S., L. Bijlsma, A. Covaci et al. 2016. Estimating community drug use through wastewater-based epidemiology. In *Assessing illicit drugs in wastewater: Advances in wastewater-based drug epidemiology*, ed. S. Castiglioni, 17–33. Luxembourg: Publications Office of the European Union.
Castiglioni, S., A. Borsotti, F. Riva, and E. Zuccato. 2016. Illicit drug consumption estimated by wastewater analysis in different districts of Milan: A case study. *Drug and Alcohol Review* 35:128–32.
Castiglioni, S., I. Senta, A. Borsotti, E. Davoli, and E. Zuccato. 2015. A novel approach for monitoring tobacco use in local communities by wastewater analysis. *Tobacco Control* 24:38–42.
Castiglioni, S., K. V. Thomas, B. Kasprzyk-Hordern, L. Vandam, and P. Griffiths. 2014. Testing wastewater to detect illicit drugs: State of the art, potential and research needs. *Science of the Total Environment* 487:613–20.

Castiglioni, S., and L. Vandam. 2016. A global overview of wastewater-based epidemiology. In *Assessing illicit drugs in wastewater: Advances in wastewater-based drug epidemiology*, ed. S. Castiglioni, 45–54. Luxembourg: Publications Office of the European Union.

Castiglioni, S., L. Vandam, and P. Griffiths. 2016. Conclusions and final remarks. In *Assessing illicit drugs in wastewater: Advances in wastewater-based drug epidemiology*, ed. S. Castiglioni, 75–76. Luxembourg: Publications Office of the European Union.

Castiglioni, S., E. Zuccato, K. Thomas, and M. Reid. 2016. Integrating wastewater analysis with conventional approaches for measuring illicit drug use. In *Assessing illicit drugs in wastewater: Advances in wastewater-based drug epidemiology*, ed. S. Castiglioni, 67–73. Luxembourg: Publications Office of the European Union.

Chiaia, A. C., C. J. Banta-Green, and J. Field. 2008. Eliminating solid phase extraction with large-volume injection LC/MS/MS: Analysis of illicit and legal drugs and human urine indicators in U.S. wastewaters. *Environmental Science and Technology* 42:8841–48.

CNN.com. 2005. Shock over cocaine river residues. https://edition.cnn.com/2005/TECH/science/08/05/italy.cocaine/index.html.

Daughton, C. G. 2001. Illicit drugs in municipal sewage: Proposed new nonintrusive tool to heighten public awareness of societal use of illicit-abused drugs and their potential for ecological consequences. In *Pharmaceuticals and personal care products in the environment: Scientific and regulatory issues*, ed. C. G. Daughton, and T. L. Jones-Lepp, 348–64. Washington, DC: American Chemical Society.

Daughton, C. G. 2003. Cradle-to-cradle stewardship of drugs for minimizing their environmental disposition while promoting human health. II. Drug disposal, waste reduction, and future directions. *Environmental Health Perspectives* 111:775–85.

Daughton, C. G. 2018. Monitoring wastewater for assessing community health: Sewage Chemical-Information Mining (SCIM). *Science of The Total Environment* 619–620:748–64.

Dell'Amore, C. 2009. Cocaine, spices, hormones found in drinking water. Accessed 12 November 2009. www.nationalgeographic.com/environment/2009/11/drinking-water-cocaine-environment/.

Down, S. 2005. Recreational drugs down the drain. 10 August 2005. www.chemistryworld.com/news/10-august-2005-recreational-drugs-down-the-drain/3000021.article.

Emke, E., S. Evans, B. Kasprzyk-Hordern, and P. de Voogt. 2014. Enantiomer profiling of high loads of amphetamine and MDMA in communal sewage: A Dutch perspective. *Science of the Total Environment* 487:666–72.

European Monitoring Centre for Drugs and Drug Addiction. 2016. *Assessing illicit drugs in wastewater: Advances in wastewater-based drug epidemiology*. Luxembourg: Publications Office of the European Union.

European Monitoring Centre for Drugs and Drug Addiction. 2018. *Perspectives on drugs. Wastewater analysis and drugs: A European multi-city study*. Insights 22. Luxembourg: Publications Office of the European Union. www.emcdda.europa.eu/topics/pods/waste-water-analysis.

Fry C., and W.D. Hall. 2004. *Ethical challenges in drug epidemiology: Issues, priorities, principles and guidelines. The GAP Toolkit Module 7.* Vienna: United Nations Office on Drugs and Crime.

Gao, J., J. O'Brien, P. Du et al. 2016. Measuring selected PPCPs in wastewater to estimate the population in different cities in China. *Science of the Total Environment* 568:164–70.

Garrison, A. W., J. D. Pope, and F. R. Allen. 1976. GC/MA analysis of organic compounds in domestic wastewater. In *Identification and analysis of organic pollutants in water*, ed. L. H. Keith, 517–66. Ann Arbor: Ann Arbor Science Publishers.

Gerrity, D., M. J. Benotti, D. J. Reckhow, and S. A. Snyder. 2011. Pharmaceuticals and endocrine disrupting compounds in drinking water. In *Biophysico-chemical processes of anthropogenic organic compounds in environmental systems*, ed. B Xing, N. Sensei and P. M. Huang, 233–49. Hoboken: John Wiley & Sons, Inc.

González-Mariño, I., R. Rodil, I. Barrio, R. Cela, and J. B. Quintana. 2017. Wastewater-based epidemiology as a new tool for estimating population exposure to phthalate plasticizers. *Environmental Science and Technology* 51:3902–10.

Gracia-Lor, E., S. Castiglioni, R. Bade et al. 2017. Measuring biomarkers in wastewater as a new source of epidemiological information: Current state and future perspectives. *Environment International* 99:131–50.

Hall, W., J. Prichard, P. Kirkbride et al. 2012. An analysis of ethical issues in using wastewater analysis to monitor illicit drug use. *Addiction* 107:1767–73.

Hart, O. E., and R. U. Halden. 2020. Simulated 2017 nationwide sampling at 13,940 major US sewage treatment plants to assess seasonal population bias in wastewater-based epidemiology. *Science of The Total Environment* 727: 138406.

Hignite, C., and D. L. Azarnoff. 1977. Drugs and drug metabolites as environmental contaminants: Chlorophenoxyisobutyrate and salicylic acid in sewage water effluent. *Life Sciences* 20:337–41.

Houston Chronicle. 2005. Italian scientists use river to gauge cocaine use. www.chron.com/news/nation-world/article/Italian-scientists-use-river-to-gauge-cocaine-use-1646310.php.

Irvine, R. J., C. Kostakis, P. D. Felgate, E. J. Jaehne, C. Chen, and J. M. White. 2011. Population drug use in Australia: A wastewater analysis. *Forensic Science International* 210:69–73.

Jones-Lepp, T. L., D. A. Alvarez, J. D. Petty, and J. N. Huckins. 2004. Polar organic chemical integrative sampling and liquid chromatography–electrospray/ion-trap mass spectrometry for assessing selected prescription and illicit drugs in treated sewage effluents. *Archives of Environmental Contamination and Toxicology* 47:427–39.

Kasprzyk-Hordern, B., and D. R. Baker. 2012. Estimation of community-wide drugs use via stereoselective profiling of sewage. *Science of the Total Environment* 423: 142–50.

Khan, M. T., M. Busch, V. G. Molina, A. Emwas, C. Aubry, and J. Croue. 2014. How different is the composition of the fouling layer of wastewater reuse and seawater desalination RO membranes?. *Water Research* 59:271–82.

Khan, U., and J. A. Nicell. 2011. Refined sewer epidemiology mass balances and their application to heroin, cocaine and ecstasy. *Environment International* 37:1236–52.

Lai, F. Y., S. Anuj, R. Bruno et al. 2015. Systematic and day-to-day effects of chemical-derived population estimates on wastewater-based drug epidemiology. *Environmental Science and Technology* 49:999–1008.

Lai, F. Y., C. Gartner, W. Hall et al. 2018. Measuring spatial and temporal trends of nicotine and alcohol consumption in Australia using wastewater-based epidemiology. *Addiction* 113:1127–36.

Lai. F. Y., J. W. O'Brien, P. K. Thai et al. 2016. Cocaine, MDMA and methamphetamine residues in wastewater: Consumption trends (2009–2015) in South East Queensland, Australia. *Science of The Total Environment* 568:803–9.

Lai, F. Y., C. Ort, C. Gartner et al. 2011. Refining the estimation of illicit drug consumptions from wastewater analysis: Co-analysis of prescription pharmaceuticals and uncertainty assessment. *Water Research* 45:4437–48.

Lai, F. Y., P. K. Thai, J. O'Brien et al. 2013. Using quantitative wastewater analysis to measure daily usage of conventional and emerging illicit drugs at an annual music festival. *Drug and Alcohol Review* 32:594–602.

Lin, A. Y., C. Lin, Y. Tsai, et al. 2010. Fate of selected pharmaceuticals and personal care products after secondary wastewater treatment processes in Taiwan. *Water Science and Technology* 62:2450–458.

Lopes, A., N. Silva, M. R. Bronze, J. Ferreira, and J. Morais. 2014. Analysis of cocaine and nicotine metabolites in wastewater by liquid chromatography–tandem mass spectrometry. Cross abuse index patterns on a major community. *Science of The Total Environment* 487:673–80.

Mastroianni, N., E. López-García, C. Postigo, D. Barceló, and M. López de Alda. 2017. Five-year monitoring of 19 illicit and legal substances of abuse at the inlet of a wastewater treatment plant in Barcelona (NE Spain) and estimation of drug consumption patterns and trends. *Science of the Total Environment* 609:916–26.

McCall, A.-K., R. Bade, J. Kinyua, F.Y. Lai, P.K. Thai, A. Covaci, L. Bijlsma, A.L.N. van Nuijs, and C. Ort. 2015. Critical review on the stability of illicit drugs in sewers and wastewater samples. *Water Research* 88: 14.

Mirror. 2005. A river of cocaine. www.mirror.co.uk/news/uk-news/a-river-of-cocaine-552689.

Mounteney, J., P. Griffiths, R. Sedefov, A. Noor, J. Vicente, and R. Simon. 2015. The drug situation in Europe: An overview of data available on illicit drugs and new psychoactive substances from European monitoring in 2015. *Addiction* 111:34–48.

National Drug and Alcohol Research Centre. 2017. *Medicine/National Drug and Alcohol Research Centre Annual Report 2016*. Sydney: University of New South Wales.

National Health and Medical Research Council, Australian Research Council, and Universities Australia. 2017. *National Statement on Ethical Conduct in Human Research 2007 (Updated 2018)*. Canberra: Commonwealth of Australia.

O'Brien, J. W, P. K. Thai, G. Eaglesham et al. 2014. A model to estimate the population contributing to the wastewater using samples collected on census day. *Environmental Science and Technology* 48:517–25.

Ort, C., M. G. Lawrence, J. Reungoat, and J. F. Mueller. 2010a. Sampling for PPCPs in wastewater systems: Comparison of different sampling modes and optimization strategies. *Environmental Science and Technology* 44:6289–96.

Ort, C., M. G. Lawrence, J. Rieckermann, and A. Joss. 2010. Sampling for pharmaceuticals and personal care products (PPCPs) and illicit drugs in wastewater systems: Are your conclusions valid? A critical review. *Environmental Science and Technology* 44:6024–35.

Ort, C., A. L. N. van Nuijs, J. Berset et al. 2014. Spatial differences and temporal changes in illicit drug use in Europe quantified by wastewater analysis. *Addiction* 109:1338–52.

Osemwengie, L. I., and S. Steinberg. 2001. On-site solid-phase extraction and laboratory analysis of ultra-trace synthetic musks in municipal sewage effluent using gas chromatography–mass spectrometry in the full-scan mode. *Journal of Chromatography A* 932:107–18.

Potter, K. W. 2006. Small-scale, spatially distributed water management practices: Implications for research in the hydrologic sciences. *Water Resources Research* 42:W03S08.

Prichard, J., W. Hall, P. de Voogt, and E. Zuccato. 2014. Sewage epidemiology and illicit drug research: The development of ethical research guidelines. *Science of the Total Environment* 472:550–55.

Prichard, J., W. Hall, E. Zuccato et al. 2016. *Ethical research guidelines for sewage epidemiology*. www.emcdda.europa.eu/system/files/attachments/10405/WBEethical-guidelines-v1.0- 03.2016%20.pdf

Prichard, J., F. Y. Lai, P. Kirkbride et al. 2012. Measuring drug use patterns in Queensland through wastewater analysis. *Trends and Issues in Crime and Criminal Justice* 442:1–9.

Prichard, J., F. Y. Lai, J. O'Brien et al. 2018. 'Ice rushes', data shadows and methylamphetamineuse in rural towns: Wastewater analysis. *Current Issues in Criminal Justice* 29:195–208.

Prichard, J., F. Y. Lai, E. van Dyken et al. 2017. Wastewater analysis for estimating substance use: Implications for law, policy and research. *Journal of Law and Medicine* 24:837–49.

Prichard, J., C. Ort, R. Bruno et al. 2010. Developing a method for site-specific wastewater analysis: Implications for prisons and other agencies with an interest in illicit drug use. *Journal of Law, Information and Science* 20:15–27.

Prinzleve, M., C. Haasen, H. Zurhold et al. 2004. Cocaine use in Europe – A multicentre study: Patterns of use in different groups. *European Addiction Research* 10:147–55.

Radford, T. 2005. River tests expose cocaine use. *The Guardian*. www.theguardian.com/world/2005/aug/05/highereducation.research.

Reid, D. E., B. J. Ferguson, S. Hayashi, Y. Lin, and P. M. Gresshoff. 2011. Molecular mechanisms controlling legume autoregulation of nodulation. *Annals of Botany* 108:789–95.

Reid, M., and K. Thomas. 2016. New psychoactive substances: Analysis and site-specific testing. In *Assessing illicit drugs in wastewater: Advances in wastewater-based drug epidemiology*, ed. S. Castiglioni, 57–65. Luxembourg: Publications Office of the European Union.

Repubblica.it., La. 2005. Scoperte tracce di cocaina nel Po 'Scorrono quattro chili al giorno'. www.repubblica.it/2005/h/sezioni/cronaca/cocainapo/cocainapo/cocainapo.html

Rodríguez-Álvarez, T., I. Racamonde, I. González-Mariño et al. 2015. Alcohol and cocaine co-consumption in two European cities assessed by wastewater analysis. *Science of The Total Environment* 536:91–98.

Rousis, N. I., E. Zuccato, and S. Castiglioni. 2016. Monitoring population exposure to pesticides based on liquid chromatography-tandem mass spectrometry measurement of their urinary metabolites in urban wastewater: A novel biomonitoring approach. *Science of The Total Environment* 571:1349–57.

Rousis, N. I., E. Zuccato, and S. Castiglioni. 2017. Wastewater-based epidemiology to assess human exposure to pyrethroid pesticides. *Environment International* 99:213–20.

Ryu, Y., D. Barceló, L. P. Barron et al. 2016a. Comparative measurement and quantitative risk assessment of alcohol consumption through wastewater-based epidemiology: An international study in 20 cities. *Science of The Total Environment* 565:977–83.

Ryu, Y., E. Gracia-Lor, R. Bade et al. 2016b. Increased levels of the oxidative stress biomarker 8-iso-prostaglandin $F_{2\alpha}$ in wastewater associated with tobacco use. *Scientific Reports* 6: 39055.

Scotsman, The. 2005. Italian rivers reveal level of cocaine use. Accessed 6 March 2018. www.scotsman.com/news/world/italian-rivers-reveal-level-of-cocaine-use-1-727304.

Sewage Analysis Core group Europe (SCORE). 2016. Ethical research guidelines for wastewater-based epidemiology and related fields. Accessed 1 March 2018. www.emcdda.europa.eu/drugs-library/ethical-research-guidelines-wastewater-based-epidemiology-and-related-fields_ru.

Ternes, T. A. 1998. Occurrence of drugs in German sewage treatment plants and rivers. *Water Research* 32:3245–60.

Thai, P. K., G. Jiang, W. Gernjak, Z. Yuan, F. Y. Lai, J. F. Mueller. 2014a. Effects of sewer conditions on the degradation of selected illicit drug residues in wastewater. *Water Research* 48: 538–47.

Thai, P. K., F. Y. Lai, M. Edirisinghe, W. Hall, R. Bruno, J. W. O'Brien, J. Prichard, P. Kirkbride, and J. F. Mueller. 2016. Monitoring temporal changes in use of two cathinones in a large urban catchment in Queensland, Australia. *Science of The Total Environment* 545: 250–55.

Thai, P. K., J. O'Brien, G. Jiang et al. 2014b. Degradability of creatinine under sewer conditions affects its potential to be used as biomarker in sewage epidemiology. *Water Research* 55:272–79.

Thomas, K. V., A. Amador, J. A. Baz-Lomba, and M. Reid. 2017. Use of mobile device data to better estimate dynamic population size for wastewater-based epidemiology. *Environmental Science and Technology* 51:11363–70.

Thomas, K. V., L. Bijlsma, S. Castiglioni et al. 2012. Comparing illicit drug use in European cities through sewage analysis. *Science of the Total Environment International* 432:432–39.

Tscharke, B. J., C. Chen, J. P. Gerber, and J. M. White. 2016. Temporal trends in drug use in Adelaide, South Australia by wastewater analysis. *Science of The Total Environment* 565:384–91.

Turque, B. 2006. Sewage tested for signs of cocaine. Accessed 6 March 2018. www.washingtonpost.com/wp-dyn/content/article/2006/03/26/AR2006032600880.html.

van der Gouwe, D., T. M. Brunt, M. van Laar, and P. van der Pol. 2017. Purity, adulteration and price of drugs bought on-line versus off-line in the Netherlands. *Addiction* 112:640–48.

van Dyken, E., P. Thai, F. Y. Lai et al. 2014. Monitoring substance use in prisons: Assessing the potential value of wastewater analysis. *Science and Justice* 54:338–45.

van Nuijs, A. L., B. Pecceu, L. Theunis et al. 2009. Can cocaine use be evaluated through analysis of wastewater? A nation-wide approach conducted in Belgium. *Addiction* 104:734–41.

Wang, J., and L. Chu. 2016. Irradiation treatment of pharmaceutical and personal care products (PPCPs) in water and wastewater: An overview. *Radiation Physics and Chemistry* 125:56–64.

Willis, K., J. Anderson, and P. Homel. 2011. Measuring the effectiveness of drug law enforcement. *Trends and Issues in Crime and Criminal Justice* 406:1–7.

Zuccato, E., S. Castiglioni, M. Tettamanti, R. Olandese, R. Bagnati, M. Melis, and R. Fanelli. 2011. Changes in illicit drug consumption patterns in 2009 detected by wastewater analysis. *Drug and Alcohol Dependence* 118(2–3): 464–69.

Zuccato, E., C. Chiabrando, S. Castiglioni et al. 2005. Cocaine in surface waters: A new evidence-based tool to monitor community drug abuse. *Environmental Health* 4:14.

Zuccato, E., E. Gracia-Lor, N. I. Rousis et al. 2017. Illicit drug consumption in school populations measured by wastewater analysis. *Drug and Alcohol Dependence* 178:285–90.

chapter three

Macro Applications of Wastewater Analysis

International Comparisons

INTRODUCTION

This chapter discusses the actual and potential international applications of wastewater analysis (WWA) to research on illicit drugs, alcohol and tobacco. Consistent with the theme of this book, our focus is on the utility of WWA for quantitative evaluation. Given the global context set for this chapter, the discussion centres on how existing survey and other data are used to compare trends between two or more countries by the United Nations Office on Drugs and Crime (UNODC) and the World Health Organization (WHO). We limit the matter traversed in this chapter to empirical considerations: the scale of substance use globally; how this is measured; challenges in data collection; how WWA is being used currently; why its expansion might be warranted; and the main practical barriers to its expansion. While the objectives of this chapter are conceptually linked to comparative international policy analysis, we do not examine the broader literature on that topic (for more material in that regard the interested reader is referred to Ritter et al. 2016; Burris 2017).

As noted in Chapter 1, at the global level illicit drugs, alcohol and tobacco share certain similarities (1.2). First, they are governed by international legal and policy frameworks – notably the United Nations' conventions on illicit drugs,[1] the *Global Strategy to Reduce the Harmful Use of Alcohol* (WHO 2010), and the United Nations treaty entitled the *WHO Framework Convention on Tobacco Control* (WHO 2003). Second, the three categories are tackled by strategies that aim to reduce market supply and demand and the harms arising from their use. Third, the type of data that are used for evaluation research are similar (1.3): survey data, qualitative information, event data and industry data (for alcohol and tobacco as legal commodities).

Chapter 1 provided basic metrics about global consumption of psychoactive substances (1.1.2). It also highlighted some of the general shortcomings of traditional data sources in estimating illicit drug consumption. In the first section of this chapter (3.1) both these issues are explored in more detail.

Although quasi-experimental designs are not feasible in the global context, there is broad recognition of the value of comparing countries on the prevalence of consumption and substance use-related harms. One of the most important themes to emerge from this section is that the greatest challenges for evaluation research are faced in low-income and lower- and middle-income countries (LMICs), where data are either unavailable or inadequate.

Section 3.2 discusses WWA. It shows that to date the bulk of WWA studies have been conducted in high-income countries and a handful of upper-middle income countries, all of which have well-developed sewerage infrastructure. The section describes the scale of cross-country research being conducted under the aegis of the EMCDDA in Europe and explains the significance of UNODC's decision to (2017b) incorporate WWA data into its assessment of global drug markets. To date, the WHO has not used WWA to monitor population trends in alcohol or tobacco or use. The utility of WWA in comparative studies between countries is discussed.

Section 3.3 addresses the question: What are the practical challenges facing WWA studies in low- and lower-middle–income countries? It first analyses information on sewerage infrastructure in these settings. Centralised sewerage treatment plants do not service many regions but it is nonetheless likely that WWA can be accommodated in multiple wastewater treatment plants in many lower-income countries.

A WWA study of substance use in Martinique in the Caribbean is instructive (Devault et al. 2014; Devault, Lévi and Karolak 2017). The study sampled a sewage treatment system that serviced a favela (slum) in ways that raise issues that may apply to other settings in low and lower-middle income countries. The chapter concludes with consideration of sources that indicate there are plans for significant investment in sewerage infrastructure in developing countries in the coming decades.

3.1 Global Metrics on Consumption of Psychoactive Substances

This chapter begins with an overview of key market indicators for illicit drugs, alcohol and tobacco. It draws chiefly on recent publications by the UNODC and WHO to demonstrate: the importance of comparisons of countries and regions in policy analysis; the integral role of survey methods, especially in understanding prevalence of drug use; and the problem of how to deal with missing or inadequate data.

3.1.1 Illicit Drugs

The UNODC's (2017a) *World Drug Report 2017* examined the nature of the global market in illicit drugs with a wide variety of detailed data on supply, health impacts and the extent of use. Supply metrics show the magnitude of illicit drug production. For example, best estimates from satellite imagery between 1998 and 2015 suggest that the total land area used to cultivate opium poppy was lowest in 2001 (142,100 hectares) and highest in 2014 (316,700 hectares) (UNODC 2016a). Calculations of global heroin manufacturing[2] range between 327 tonnes in 2015 to a high of 686 tonnes in 2007. Best estimates of coca bush cultivation over a similar time period range between 120,800 (2013) and 181,600 (2007) hectares. The UNODC (2016a) calculated[3] that the potential output of pure cocaine exceeded 900 tonnes annually in the years 2006 to 2014.

Global seizure data for 2015 show that cannabis was confiscated in the highest volume – 7,317 tonnes, followed by cocaine (864 t), opium (587 t), methamphetamine (132 t), pharmaceutical opioids (113 t), heroin and morphine (80 t), gamma-butyrolactone (GBL), ketamine and new psychoactive substances (NPS) (57 t), amphetamine (52 t) and ecstasy (6 t) (UNODC 2017b).

Illegal tobacco and alcohol are also seized in sizeable volumes. In 2016 Port Control Units participating in the UNODC's Container Control Programme reportedly captured 405 million cigarettes and almost 6 tonnes of tobacco and 167,294 litres of alcohol (UNODC 2016b).

With respect to health impacts, in 2015 an estimated 190,000 people died prematurely from drug use (mainly attributable to opioid use) (UNODC 2017a). In the same year, approximately 29 million people were estimated to have a drug use disorder; 12 million people had injected drugs, of whom 14% were estimated to be living with HIV and 50% with hepatitis C.

Survey and other official data suggest that approximately 250 million people aged 15–64 years used at least one illicit drug in 2017 (UNODC 2017a). The most commonly used drug was cannabis, with 183 million users, followed by amphetamines and prescription stimulants (37 m), opioids (35 m), ecstasy (22 m), opiates (18 m) and cocaine (17 m).

Estimates of the percentage of population that used each drug type have been made by region (UNODC 2016a). The prevalence estimates for North America are several times higher than the Caribbean while Central Asia has higher prevalence rates of use of cannabis and opioids than East and Southeast Asia. However, data are not available on the use of other drugs for Central Asia. The same is true of the African sub-regions.

The UNODC is implementing strategies to increase capacity for illicit drug research in regions with missing or inadequate data. Notably, its *Annual Report Covering Activities During 2016* contains details of the first Regional Workshop on Collection and Analysis of Data on Drug Use and Estimation of Drug Users among the General Population, which was held in Dakar, Senegal (UNODC 2016b). This event incorporated training sessions that were facilitated by experts in drug use epidemiology from the UNODC, the Inter-American Drug Abuse Control Commission/ Organization of American States and the South African Community Epidemiology Network on Drug Use (Medical Research Council).

Importantly, World Drug Report data show that drug use is not a phenomenon peculiar to high-income regions. For instance, globally the region with the highest estimated prevalence of cannabis use was West and Central Africa (12.4%). The figures for opioids and cocaine use in the same region were 0.44% and 0.7% but these estimates equate to about 1 million opioid users and 1.7 million cocaine users in West and Central Africa. These data are comparable to those provided for Western and Central Europe: 0.5% for opioids and 1.1% for cocaine. The UNODC (2016b) has expressed concern about the growing demand for heroin in Africa, which appears to be a major transit point for opiates trafficked from Afghanistan to Europe and North America.

3.1.2 Alcohol

The WHO monitors alcohol consumption primarily through the Global Information System on Alcohol and Health (GISAH).[4] This collates data from WHO Member States, national and international surveys, global burden of disease projects, and industry (see WHO 2014, 26, 346). GISAH estimates that in 2016 6.4 litres of pure alcohol were consumed per capita by those aged 15 years and older (WHO 2017a).

There is considerable variation between countries – ranging from 100 ml in Libya to 18.2 litres in Lithuania. Many high-income countries are above the global average, including Belgium (13.2 litres), the UK (12.3 litres), Germany (11.4 litres), Australia (11.2 litres), Switzerland (10 litres) and the US (9.3 litres). However, comparable estimates are found in low- and lower-middle–income countries with 10–12 litres consumed in Angola, Rwanda, Equatorial Guinea and Uganda.

3.1.3 Tobacco

Survey data are also integral to the WHO's (2017b) efforts to monitor global tobacco consumption and exposure to tobacco smoke. This is done primarily through national and regional surveys and incorporates the WHO's (2017b) projects, the Global Youth Tobacco Survey (GYTS) and the Global Adult Tobacco Survey (GATS). Additional information comes from longitudinal studies (e.g. Doll et al. 2004) and the International Tobacco Control Policy Evaluation Project.[5]

There are full-scale survey-centred monitoring systems in 76 countries (WHO 2017b). A further 39 countries have recent representative-survey data for adults and youths. If countries ensure that these surveys are repeated every five years, the WHO will accept their datasets as an adequate form of monitoring (WHO 2017b). The major resources required for survey-centred monitoring means that more than half of the low-income countries either have weak tobacco monitoring, or none at all. The comparable figures for high and middle-income countries respectively are 5% and 15% (WHO 2017b).

Additional data are sourced from domestic government records on health, tobacco imports and tobacco taxation. Worldwide there were estimated to be 1.1 billion smokers in 2015 (WHO 2017b), 80% of whom live in low- and middle-income countries (USNCI and WHO 2016).

3.1.4 *Recent Use*

Information on recent use of psychoactive substances is gathered by national and regional surveys. By way of example, the most recent US *National Survey on Drug Use and Health* (SAMHSA 2017) recruited about 68,000 participants aged 12 years and older and estimated that 50.7% of the American population were current users of alcohol, as indicated by use in the past month. The figure for tobacco was 23.5%. Approximately 10.6% of the population were current users of an illicit drug. This included cannabis (8.9%), cocaine (0.7%), hallucinogens (0.5%), methamphetamine (0.2%) and heroin (0.2%).

3.2 *Evaluation Research at the International Level*

Chapter 1 discussed two aspects of evaluation research: data generation (through surveys, qualitative data, event data and industry data) and research designs (descriptive analyses, randomised controlled experiments and quasi-experimental designs) (see 1.3). This chapter discusses special issues and complexities in conducting international evaluation research to prepare for a critical analysis of the ways in which WWA might be able to complement such research.

3.2.1 *Research Designs*

Randomised controlled trials provide a means by which researchers can attempt to isolate the causal effect of variables of key interest. But, they are very difficult to implement at the national level (Babor et al. 2010b; Hall 2018) and are an impractical way of making comparisons between countries. Quasi-experimental national designs are feasible when changes are made to policy or when circumstances independently affect markets for psychoactive substances (Babor et al. 2010a; 2010b). Yet, their main limitation lies in their inability to control for the effects of extraneous factors. Additionally, when they are attempted at a national level, challenges arise if analyses are conducted retrospectively using data that are inadequate because they were collected for other purposes (Hall 2018).

Problems are multiplied when quasi-experimental studies are attempted at the international level – for example by attempting to assess the impact of a new policy in one country by comparing its effects on drug use with that in other countries where the policy was not implemented. Among other things, the large numbers of variables on which the countries may differ complicates international quasi-experimental studies.

Consequently, the main method of evaluation undertaken at the international level is often a comparison of descriptive statistics on drug use between countries. It is important not to underestimate the value of comparative analyses. While descriptive comparisons are limited in their capacity to confidently *evaluate* policy, there is no doubt that descriptive comparisons influence policy agendas, posing examples of 'better performance' for policy formation, and potentially sway political decision making. As the UNODC and WHO data illustrated in the previous section, descriptive trend data provide a global picture of markets in psychoactive substances. Without these, international agencies would have no means of identifying global priorities; national governments would lack metrics on which to compare and monitor the 'performance' of their country in terms of drug consumption prevalence, substance use-related harms and the like.

3.2.2 Data Generation

Every method of collecting data on psychoactive substances has strengths and weaknesses, as was discussed at the end of the last chapter. This includes WWA (see 2.5). But for international comparative analyses there are additional overarching problems. Data are not collected in some jurisdictions, or are collected using inadequate methods. This problem, which is discussed further below, is particularly acute in low- and lower-middle–income countries.

However, another enduring problem complicates comparisons between even high-income countries with well-established systems for monitoring psychoactive substances. As highlighted by Kilmer, Reuter and Giommoni's (2015) analysis, most survey data in most countries are not comparable because they are collected with different procedures, report different measures or sample different age groups.

3.2.2.1 Surveys

In *surveys*, the main categories of procedures include face-to-face interviews, telephone surveys and mail surveys. There are further variations within each of these categories, for instance as to whether they make use of a computer, audio technology, a human interviewer, or – in the case of mail surveys – whether participants' completed mail surveys are collected from their residences or returned by the participants through the postal system (Kilmer, Reuter and Giommoni 2015).

Literature on 'social desirability bias' suggests that survey participants may (wittingly or unwittingly) report information that presents themselves in a positive light (e.g. Harrison and Hughes 1997). This phenomenon may operate even when participants are able to provide information anonymously on topics that are not particularly sensitive or embarrassing. With this background, it is not surprising that the results of surveys about drug use – which is a topic that can attract comparatively high social stigma – appear to differ according to the degree of anonymity they afford participants. It is also known that these procedures differ in their capacity to recruit samples of participants who are representative of the general population (see e.g. Decorte et al. 2009; Johnson 2015; Kilmer, Reuter and Giommoni 2015).

These issues mean that international differences between countries in the reported prevalence of drug use may partly be an artefact of the procedures employed to measure it. Recent empirical analyses seem to support this point. Giommoni, Reuter and Kilmer (2017) compared the national population surveys of the US, Canada, Europe and Australia. Prevalence estimates were adjusted in an attempt to mimic the use of a standard data collection procedure across all countries. The results suggested that survey procedures did affect cross-country comparisons to a degree.

Problems may be further compounded when survey participants differ in age between different countries. Kilmer, Reuter and Giommoni (2015) argue that *broader* age ranges reduce the prevalence of drug use within a population because they incorporate a greater proportion of participants who are unlikely to use drugs – namely those aged between 12 and 14 years of age and those over 64 years.

Trend data are critical in most comparisons of drug use between countries. Two issues impair the value of trend data. First, the frequency of sampling varies. For instance, between 2000 and 2010 the frequency of sampling in some high-income jurisdictions ranged from annual surveys (e.g. US, England and Wales), to every three (e.g. Germany, Australia), four (e.g. Netherlands) and five years (e.g. France) (Kilmer, Reuter and Giommoni 2015). This means that an examination of trends over one decade might have ten data points for one country and only two for another.

Second, countries alter their sampling procedures in ways that alter the apparent prevalence of drug use. For instance, Kilmer, Reuter and Giommoni (2015) suggest that a reduction in reported cannabis use in Germany between 2003 and 2006 from 6.9% to 4.7% may have reflected a change in the age of participants and the questions employed.

The problems in comparing national surveys are off-set to some degree by using data from large surveys conducted using comparable methods in multiple countries. By taking careful measures to ensure that procedures are consistent across countries and by recruiting participants of the same age range, these surveys provide greater confidence in data used for comparisons. The features of leading international surveys are presented in Table 3.1.

Three of the four representative international surveys focus on young people and children, namely the European School Survey Project on Alcohol and Other Drugs (ESPAD), WHO Health Behaviour in School-Aged Children (HBSC) and International Self-Report Delinquency Study (ISRD). This age cohort is easier to recruit and an important one to study because of its vulnerability to impaired physical, psychological, social and educational wellbeing. Trends observed in these surveys are valuable for comparative purposes but their utility for estimating prevalence of use in the general population is limited because they exclude adults.

The WHO World Mental Health Survey (WMHS) employs a very robust and resource-intensive procedure to train its interviewers and gather data. It has a particular focus on the most widely used drugs, namely, alcohol, cannabis and tobacco, and it provides information on prevalence of use, dependence on alcohol and drugs, and therapeutic needs (see Degenhardt et al. 2010).

Because substance use is relevant to different domains of life, the HBSC approaches the topic from the perspective of health, the WMHS from mental health. The ISRD adopts the standpoint of crime victimisation and offending (Enzmann et al. 2018). ESPAD examines psychoactive substance use in detail by asking participants about their use of a wide variety of substances.

In terms of use for monitoring purposes, ESPAD and HBSC produce data every four years. The ISRD has been conducted in three waves and to date the WHMS has been conducted once in each of the participating countries. None of the surveys is conducted in countries defined as low-income countries by the World Bank (2017).

3.2.2.2 Event Data

Chapter 1 shows that a wide variety of official statistics (event data) are generated by government agencies whose activities are related to psychoactive substances (see 1.3). One of the key challenges in interpreting these data at a local or national level is that it is difficult to determine the extent to which changes in the data reflect changes in agencies' practices or changes in drug market activity, including consumption levels.

International comparative analyses of event data are difficult because countries do not record event data in the same ways. To summarise some

Table 3.1 Characteristics of Multi-country Representative Surveys which Collect Data on Psychoactive Substances

	Time Frame Between Sampling	Participants' Age (years)	Countries (N)	Low-Income Countries* (N)	Focus
European School Survey Project on Alcohol and Other Drugs[a] (ESPAD)	4 years	15–16	35	N/A	Substance use
WHO Health Behaviour in School-Aged Children[b] (HBSC)	4 years	11, 13, 15	48	–	Health (alcohol, tobacco, cannabis)
International Self-Report Delinquency Study[c] (ISRD)	Periodic**	12–16	35	–	Crime victimisation & offending (alcohol & drugs)
WHO World Mental Health Survey[d] (WMHS)	Single samples***	>17	25	–	Mental health (alcohol, drugs, nicotine)

* Based on World Bank 2017 listing.
** 1991–1992, 2006–2008, 2012–present
*** The survey has been conducted once in each of the 25 countries, beginning in 2001 in Belgium, France, Italy, Mexico, China (Beijing, Shanghai), Spain and the US. The most recent data collection phase occurred in Argentina (2015).
[a] European School Survey Project on Alcohol and Other Drugs. www.espad.org/.
[b] Health Behaviour in School-Aged Children. About HBSC. www.hbsc.org/about/index.html. For participating countries see Health Behaviour in School-Aged Children. HBSC member countries. www.hbsc.org/membership/countries/index.html.
[c] Northeastern University. The International Self-Report Delinquency Study. https://web.northeastern.edu/isrd/.
[d] Harvard Medical School. The World Mental Health Survey Initiative. https://www.hcp.med.harvard.edu/wmh/. For participating countries see Harvard Medical School. WMH Cross National Sample. www.hcp.med.harvard.edu/wmh/national_sample.php.

key arguments mounted by Kilmer, Reuter and Giommoni (2015) primarily about drug use in high-income countries:

(a) Estimates of numbers of people using drugs in problematic ways are produced infrequently, using different methods applied to administrative data often defined in dissimilar ways.
(b) Cross-country variations in the production of death certificates muddies comparisons of countries' drug-related deaths.
(c) Drug-related arrests are complicated because the meaning of arrest varies across criminal justice systems, arrests are recorded differently, they are linked to law enforcement operational priorities, and criminal offences are defined inconsistently.
(d) Country seizure data may differ depending on the extent to which countries are drug producers, transit points, or consumers. For example, a seizure of heroin in North America would better reflect domestic consumption than would one in Africa (a transit point) or Afghanistan (a producer) (see UNODC 2016b). Disparities exist between countries in terms of the skill of the law enforcement agencies in interdiction, the care taken by traffickers to avoid detection, the operational priorities of the agencies, and whether the purity of seizures is tested and reported.
(e) Analysis is complicated even in the few countries that routinely collect and report price and purity data. For example, an increase in price may reflect a shortage of supply and hence reduced consumption. It also could mean that consumption is increasing and suppliers are confident that the market will accept price increases. In regards to purity data, there is a wide range in practices in forensic laboratories, which is usually determined by jurisdictional legal and/or workload influences. For example, some jurisdictions may opt to measure purity only for seizures that exceed prima facie limits, such as those that differentiate simple possession from possession for sale. Currently, it is impractical to use price and purity data for international comparisons.

3.2.3 *Special Challenges in Measuring Alcohol and Tobacco Use*

Comparative analyses of alcohol and tobacco surveys face similar problems of missing data, inadequate data, infrequent data collection, inconsistent data collection procedures and so forth. But there are other issues that are unique to alcohol and tobacco, respectively.

3.2.3.1 Unrecorded Alcohol Consumption

Industry data, including taxation and trade statistics, are used in many countries to estimate per capita alcohol consumption per year. These industry data cover 'recorded consumption' but significant amounts of the alcohol consumed in each country may not be recorded. This may occur because alcohol is brewed at home. Unrecorded alcohol may also be brewed locally for profit in an unregulated (and potentially illegal) way and then trafficked, or sold to cross-border shoppers, or diverted from industrial or medical sources (WHO 2014). Estimates of 'unrecorded consumption' are difficult to make and differ between countries in ways that probably reflect the affordability of recorded alcohol. Estimates by the WHO (2014) suggest that globally one-quarter of alcohol consumption is unrecorded. Proportionately about 8.5% of alcohol is unrecorded in high-income countries and the figure for low- and lower-middle–income countries is about 40%. In some Islamic states where alcohol is banned nearly 100% of alcohol consumption is unrecorded (WHO 2014). The WHO (2014, 84) estimates of unrecorded alcohol consumption are made in different ways, including direct metrics from national surveys, expert opinion, indirect estimates from government data on confiscated or seized alcohol, and indirect estimates from survey data.

The WHO (2014) has estimated unrecorded alcohol consumption for some time because a failure to do so would mean that global patterns of alcohol use are incomplete (Babor et al. 2010b). The potential adverse health implications of unrecorded alcohol have been highlighted by Rehm and colleagues (Rehm et al. 2014; Rehm, Kanteres and Lackenmeier 2010). The lower price of unrecorded alcohol means that it is disproportionately consumed by people from lower socio-economic backgrounds (in any country) and often in larger amounts than recorded alcohol. The consumption of unrecorded alcohol may also contribute more to the global disease burden because it may contain toxic metals, alcohol congeners (e.g. acetaldehyde), carcinogens, or highly toxic industrial alcohol that has been made unfit for human consumption (e.g. denatured alcohol, which may contain methanol).

3.2.3.2 Tobacco Monitoring

The tobacco industry may provide data in some countries, but globally the industry is seen to hamper rather than assist in monitoring tobacco use. Trends in tobacco use in survey data clearly show that global tobacco reduction strategies have successfully reduced global consumption to a degree not matched by strategies that have targeted alcohol or illicit drugs (USNCI and WHO 2016). Global figures indicate a 2.8% reduction in the rates of smoking among those aged 15 years and older, from 23.5% in 2007 to 20.7% in 2015 (WHO 2017b). The decline has been largest in high-income countries. Reductions were observed in about half of the middle-income countries and a third of the low-income countries.

Nonetheless, the scale of tobacco use is still very large and on current trajectories it may be difficult for WHO Member States to reach their 2025 target of a 30% reduction (USNCI and WHO 2016). Although the percentage of people who smoke declined between 2007 and 2015, the absolute numbers of smokers remained stable at about 1.1 billion people because of population growth (WHO 2017b). The effects of the disease burden attributable to tobacco is unevenly distributed between countries; 80% of adult men who smoke cigarettes currently live in low- and middle-income countries – "foreshadowing grave consequences for health in these countries" (USNCI and WHO 2016, xv).

The WHO (2017b) strongly emphasises the importance of maintaining and, if possible, improving tobacco monitoring. Between 2007 and 2014 the WHO recorded a steady increase from 46 to 77 in the number of countries using best-practice monitoring standards. The figure for 2016 was 76, suggesting a plateau in improvements to monitoring systems. This is partly driven by the difficulties that low-income (and some middle-income) countries face in conducting regular representative surveys of smoking in the adult and youth populations (see WHO 2017b).

The WHO (2017b) also identified a deficiency in data collection on the breadth of tobacco products. Notably, in the last decade only a third of countries worldwide have collected and reported survey data relating to tobacco use that includes not only cigarettes, but other forms of smoked tobacco (e.g. pipes, bidis, roll-your-own) and smokeless tobacco[6] products, such as snuff, snus, gutka and chewing tobacco. While improvements have been made, particularly in youth surveys that adopt WHO protocols, the WHO (2017b, 56) has stated that "effectively combatting the tobacco epidemic requires all types of tobacco use to be monitored in all countries."

3.3 *Current Use of Wastewater Analysis at the International Level*

How is WWA currently being used in the international arena and particularly for comparative analyses? At the time of writing, WWA has been undertaken in 38 countries. Table 3.2 lists those countries and provides a sample citation for each setting. The World Bank's (2017) categorisation of national economic income status is included (low, lower-middle, middle, upper-middle, high).

Table 3.2 shows that most of the WAA studies have been undertaken in 38 European countries classified as high-income by the World Bank. Two regions of France, namely Martinique and Reunion Island, represent the only WWA studies conducted in the Caribbean and the Indian Ocean, respectively. The discussion will return to the significance of the Martinique study shortly.

Table 3.2 Wastewater Research by Country, Income Status and Sample Study

#	Country	Income Status	Sample Citation
1	Australia	High	Lai et al. 2013
2	Austria	High	EMCDDA 2018
3	Belgium	High	EMCDDA 2018
4	Bosnia & Herzegovina	Upper-middle	EMCDDA 2018
5	Brazil	Upper-middle	Maldaner et al. 2012
6	Canada	High	Metcalfe et al. 2010
7	China	Upper-middle	Khan et al. 2014
8	Colombia	Upper-middle	Bijlsma et al. 2016
9	Croatia	Upper-middle	Terzic, Senta and Ahel 2010
10	Cyprus	High	EMCDDA 2018
11	Czech Republic	High	Baker et al. 2012
12	Denmark	High	EMCDDA 2018
13	England	High	EMCDDA 2018
14	Finland	High	EMCDDA 2018
15	France	High	EMCDDA 2018
16	France (Martinique)	High*	Devault et al. 2014
17	France (Reunion Island)	High	Nefau et al. 2013
18	Germany	High	EMCDDA 2018
19	Greece	High	EMCDDA 2018
20	Iceland	High	EMCDDA 2018
21	Ireland	High	Bones, Thomas and Paull 2007
22	Israel	High	Gonzalez-Marino et al. 2019
23	Italy	High	EMCDDA 2018
24	Lithuania	High	EMCDDA 2018
25	Malta	High	EMCDDA 2018
26	New Zealand	High	Wilkins et al. 2018
27	Norway	High	EMCDDA 2018
28	Poland	High	EMCDDA 2018
29	Portugal	High	EMCDDA 2018
30	Romania	Upper-middle	EMCDDA 2018
31	Serbia	Upper-middle	EMCDDA 2018
32	Slovakia	High	EMCDDA 2018
33	Slovenia	High	EMCDDA 2018
34	Spain	High	EMCDDA 2018
35	South Africa	Upper-middle	Archer et al. 2018
36	South Korea	High	Kim et al. 2015
37	Sweden	High	Östman et al. 2014

(*continued*)

Table 3.2 (Cont.)

#	Country	Income Status	Sample Citation
38	Switzerland	High	EMCDDA 2018
39	The Netherlands	High	EMCDDA 2018
40	United States	High	Banta-Green et al. 2009

* Martinique is a territory of France. The gross national income per capita of Martinique was estimated to be €19,050 (USD22,866) in 2007 (see Moriame and Greliche 2007), which is above the World Bank threshold for a high-income country: USD12,236 (see World Bank 2017).

Source: adapted from EMCDDA (2016).

Six studies were carried out in upper-middle income countries, in: Bosnia and Herzegovina, Brazil, China, Colombia, Croatia, Romania, Serbia and South Africa. To date, no WWA studies have been conducted in other low-, lower-middle–, or middle-income countries.

3.3.1 *European Monitoring Centre for Drugs and Drug Addiction*

The very strong representation of European countries reflects the activities of the SCORE network, discussed in the last chapter (see 2.2.2). This study is done under the aegis of the European Monitoring Centre for Drugs and Drug Addiction (EMCDDA). EMCDDA is a unique agency that is involved in the production of comparable data sources across multiple member countries of the EU. It uses these data for annual comparative analyses (Kilmer, Reuter and Giommoni 2015). The numbers of countries and cities the EMCDDA has incorporated into WWA monitoring has expanded over the last several years. Results have been annually uploaded to the EMCDDA's website since 2011, including interactive maps representing estimated consumption levels.

The EMCDDA's publications examine a wide variety of drugs and note the capacity for WWA to estimate population consumption of alcohol and tobacco (e.g. EMCDDA 2016). However, the EMCDDA's WWA monitoring system does not produce estimates of consumption of alcohol or tobacco because the agency's mission is confined to illicit drugs.

3.3.2 *Australian National Wastewater Monitoring Program*

The government-funded Australian WWA monitoring program currently produces three reports each year (ACIC 2019). It presents data based on wastewater samples from 50 treatment sites around the country, the catchments of which collectively service an estimated 54% of the Australian population (i.e. 12.6 million people). The program measures consumption of a wide variety of psychoactive substances. This includes alcohol, tobacco, illicit drugs (e.g. methamphetamine, MDMA, cocaine, heroin, cannabis (THC), and a range of NPS). While the program was initially funded as a three-year pilot study, the Australian government recognised that the strength of the program depended on good time series data and have since funded data collection for an additional four years.

Even though it is funded by the Australian Criminal Intelligence Commission (ACIC), the ACIC CEO's Foreword in the most recent reports states "*wastewater analysis is a tool to measure and interpret drug use within populations, providing a measure of one important aspect of national health—the demand for a range of licit and illicit drugs*" (ACIC 2020, 1).

Some of the trends identified in individual reporting periods include "decreases in nicotine, alcohol, methamphetamine, cocaine, MDMA, cannabis (capital city) and fentanyl consumption between April 2019 and August 2019, and the increases in heroin, oxycodone and cannabis (regional) consumption" (ACIC 2020, 1). Highlights of time trends over the course of the program to date include identifying that "population-weighted average consumption of nicotine, methamphetamine and cocaine increased, while the consumption of alcohol, oxycodone and fentanyl decreased" (ACIC 2020, 1).

The program also estimates the total annual mass of specific substances consumed. The street value of illicit substances consumed can be estimated by triangulation with price data. Additionally, these data provide insight into the geographic areas where certain drug classes are used and information on when they are used, for example during specific events or during the week.

The Australian WWA monitoring program is evolving as substances are added into the program through policy interest or technical improvements (e.g. heroin was included from Report 3; cannabis (THC) through the measurement of THC-COOH was included from Report 6). Substances can also be removed if they prove of little interest or there are too few detections (e.g. the synthetic cannabinoids JWH-018 and JWH-that 03 were discontinued after Report 3). Additionally, the samples are archived so it is possible to analyse samples retrospectively once new methods are developed for additional compounds or higher precision estimates can be provided of existing compounds. The accuracy of the program has also improved. For example, higher accuracy population estimates were applied from Report 4 onwards and also applied retrospectively. More accurate flow data became available for some previously reported samples that were included retrospectively in Report 9.

This success of the program has prompted other nations to introduce similar programs, e.g. the USA, New Zealand, Canada, China and some

European countries, and to seek advice from the Australian program on how to do so.

3.3.3 *United Nations Office on Drugs and Crime*

It is interesting to compare the use made of WWA in recent *World Drug Reports* of the UNODC. No reference was made to WWA in the 2015 report (UNODC 2015). In 2016 the UNODC (2016a) used WWA data from SCORE, which it interpreted as evidence that cocaine use was stable in Europe between 2011 and 2014 although it differed between cities. The 2016 report also included WWA estimates of the consumption of cocaine, amphetamine, methamphetamine and cannabis in Europe, Australia and Canada.

The 2017 report showed greater interest in WWA. The report explained that it engaged with WWA because it was seen as an alternative to measuring cocaine consumption using seizure data and household surveys, both of which suffered from major limitations. The UNODC (2017a) concluded after triangulating seizure data, surveys and WWA that there was an upward trend in cocaine consumption in Europe. In other words, the agency did not simply present WWA trends. Wastewater analyses were integrated with traditional drug monitoring data to inform the UNODC's conclusion on trends in cocaine use.

A similar approach was adopted in the agency's analysis of the cocaine market in Oceania – concentrating on Australia. Triangulation in that region integrated data from: Australia's general population survey; surveys that target people who inject drugs and people who use ecstasy; drug price indices; and wastewater data derived from Australia's national WWA monitoring program. The UNODC (2017a, 35) noted the lower estimate of cocaine consumption derived from Australian wastewater analyses in comparison with European consumption estimates. Comparisons were also made to cocaine consumption in the Caribbean (Martinique), Colombia (Medellin, Bogota), Canada (Montreal), the United States (Seattle), South Korea (Busan) and New Zealand (Auckland) (UNODC 2017a, 31).

Discussion of international drug markets drawing on WWA data were also incorporated into *World Drug Reports* in 2018 (UNODC 2018) and 2019 (UNODC 2019). The fact that the UNODC has begun using WWA data in these ways is significant in the short history of the WWA field. The UNODC has demonstrated its confidence in WWA as a cross-country comparator. As best as we can determine, the UNODC is the only international agency to date to incorporate WWA data into its publications. The WHO does not appear to have used WWA estimates of consumption of alcohol or tobacco. In part that may be because research has only recently supported the validity of these methods (see Lai et al. 2018).

3.4 *The Viability of New Applications of Wastewater Analysis Globally*

This final section of the chapter draws on the discussion thus far to critically assess potential new applications of WWA in global public health. Our focus is on examining where WWA could be of most practical use to the international community.

Harking back to the limitations of WWA that were reviewed at the end of Chapter 2, WWA cannot estimate the numbers of people in a country who use psychoactive substances. So, for instance, it cannot discern whether a country has many light users or a small number of heavy users. Wastewater analysis can only give a macro-picture of the total estimated consumption per 1,000 people within a given catchment area.

Still, WWA data can efficiently monitor consumption trends within very large populations. With the support of the EMCDDA, in recent years SCORE has demonstrated that WWA is feasible and robust across multiple countries. Both the EMCDDA and the UNODC have shown how descriptive trend data from WWA studies are useful when combined with other traditional data sources on drug use.

Until recently, WWA researchers have not attempted to list the advantages of WWA for comparative international research for psychoactive substances, including alcohol and tobacco. The main apparent advantages are worth highlighting here.

1. *Annual data collection*
   As shown in Europe, WWA can be conducted annually in multiple countries, thereby circumventing the limitation of infrequent data collection in many national and most international or regional surveys of drug use.
2. *Consistent and comparable metrics between countries*
   WWA data can provide a consistent and comparable metric between countries – providing that inter-laboratory calibration tests are conducted and the same protocols are followed for sampling and analysis. This capability of WWA circumvents the problems of inconsistencies between countries in: (a) how they conduct national surveys (e.g. procedures, age definitions); and (b) how they collate event data – such as drug-related seizures, arrests and deaths, and contact with problematic substance use behaviours.
3. *Lower risk of affecting trend data through methodological changes*
   Trend data from surveys can be affected by changes in the ways surveys are conducted or the age groups that are targeted. Arguably, the WWA method is less susceptible to these changes. WWA researchers, mainly in the university sector, have an interest in

ensuring that their methods are in step with those conducted in other countries. It is true that improvements in WWA methods may affect the comparability of WWA data over time. This has occurred in 2010 when new standards were introduced for sampling and analysing wastewater.

4. *Easier to interpret than global seizure data (re consumption)*

   The degree to which drug seizure data reflects consumption in a country depends upon whether the country is a net-exporter of drugs, a net-importer or a transit point. By contrast, it is not necessary to account for a country's predominant role in the global drug trade to interpret WWA data. Furthermore, seizure data are sensitive to fluctuations in law enforcement priorities and the allocation of resources to, or away from, drug law interdiction.

5. *Easier to interpret than price and purity data (re consumption)*

   WWA data are easier to routinely report and interpret than data on the price and purity of drugs.

6. *Able to estimate total alcohol consumption ('recorded' and 'unrecorded')*

   WWA estimates of alcohol consumption (e.g. Lai et al. 2018) provide an alternative way for the WHO to account for both recorded and unrecorded alcohol consumption. Currently the WHO uses national estimates, but countries use an inconsistent mixture of approaches to account for unrecorded alcohol, including survey data, expert opinion and alcohol seizures. WWA does not need to account for cultural practices and other factors that may cause variation in the consumption of unrecorded alcohol. In short, it can potentially provide a consistent metric across countries.

7. *Able to estimate consumption of all forms of smoked tobacco (and possibly smokeless tobacco products)*

   The WHO could also use WWA to provide data on trends in tobacco consumption. Annual WWA estimates of tobacco use would markedly increase the data points that WHO can use to monitor tobacco consumption in many regions. Unlike surveys which collect data only on cigarettes, WWA data can account for all types of tobacco products.

8. *Could benefit lower-income countries that currently lack capacity to undertake monitoring with traditional methods*

   Lower-income countries, which can only afford to monitor consumption of psychoactive substances infrequently or not at all, may be able to use WWA to markedly improve data collection on illicit drugs, alcohol and tobacco. These countries are least able to bear the economic and health burdens of substance use.

   This is not to discourage efforts to build capacity in these countries to undertake monitoring with traditional methods. (See 3.1.1 regarding the capacity-building activities of the UNODC and other agencies in Africa.) As we have repeatedly stated, WWA is a supplement for traditional methods, not a replacement. It is also true that the WWA field has identified areas where the method needs to be improved – chief among them is in estimating the size of the population contributing to wastewater samples. However, in our view it is self-evident that WWA data are (a) better than no data at all and (b) especially useful where other data sources are scant.

Points 1 to 5 are probably not contentious given (a) previous analyses of the benefits of WWA for monitoring drugs (EMCDDA 2016; UNODC 2017a, 13). It is also relevant that non-WWA scholars recognise the capability for WWA to bypass limitations of survey data for international comparative research. Kilmer, Reuter and Giommoni (2015, 258) endorse arguments made by Burgard, Banta-Green and Fields (2014, 1366) that the "large error term around a valid [WWA] estimate" is preferable to "a small error term around an invalid [survey] estimate".

So far UNODC has placed greatest confidence in WWA estimates of cocaine consumption. Nonetheless, it has reported WWA data on amphetamines, methamphetamine and cannabis (UNODC 2016, 2017a). UNODC seems prepared to broaden its use of WWA data on other drugs if advances in the WWA field can reduce uncertainties in estimating the consumption of substances other than cocaine.

The situation concerning points 6 and 7 is less clear. It is difficult to discern whether the WHO is aware of the capabilities of WWA with respect to alcohol and tobacco. But as far as we are aware, this chapter is the first to argue the case for WHO considering WWA for monitoring alcohol or tobacco.

The last (eighth) point may be of particular interest to the UNODC and WHO. We think that it is probably the most important topic in this chapter. Consequently we will explore the practicality and viability of expanding WWA into lower-income countries – preferably without imposing significant strain on domestic resources.

3.4.1 Practical Challenges for Implementing Wastewater Analysis in Low-, Lower-Middle– and Middle-Income Countries

Table 3.2 listed 38 countries in which at least one WWA study has been conducted. Of these, 30 are classified as high income by the World Bank (2017) and 8 are upper-middle income countries. The expansion of the method into new countries is worth considering, including high and upper-middle income countries. A sound model for planning such an expansion has been demonstrated by SCORE.

Recent WWA studies in Colombia (Bijlsma et al. 2016) and South Africa (Archer et al. 2018) demonstrate that the expertise of WWA researchers[7] in one country can be used to assist in implementing best-practice WWA research methods in countries many thousands of kilometres away. Colombia and South Africa are both upper-middle income countries (World Bank 2017). Can similar WWA methods be implemented in lower-income countries?

Chapter 2 described how WWA uses public sewerage infrastructure for data collection. The 'data' of interest literally pool in wastewater treatment plants (WWTPs). Several points made in Section 2.2.2 are worth recapping. Wastewater authorities are important stakeholders in WWA research because they can facilitate (or even take carriage of) sampling. These authorities also have essential information about the sewerage infrastructure under their control e.g. whether it collects storm water or industrial waste; the approximate size of the population the system serves; data about the degree to which wastewater leaks from the system; and whether wastewater from other catchments is pumped or transported into to the system. These basic questions need to be answered to assess the viability of using WWA in lower-income countries. Do the countries in question have sewerage infrastructure suitable for WWA sampling? What is known about the infrastructure? And does climate present any novel technical hurdles?

For lower-income countries several potential challenges may impede WWA. Perhaps the greatest of these is absence of sewerage infrastructure.

3.4.1.1 Inadequate Sewerage Infrastructure

The United Nations and other agencies aim to improve the global management of wastewater because of its major role in protecting human health and reducing environmental degradation. For these reasons, different approaches have been taken to assess the extent and quality of sewerage infrastructure around the world.

Global estimates suggest that about 60% of the world's population is connected to a sewer system (WWAP 2017). However, the lack of publicly available information on sewerage impedes the development of metrics in many regions. For instance, Sato and colleagues (2013) searched for information on wastewater management in 181 countries. They found that in 57 countries no information was "available on any aspect of wastewater production, treatment, or use" (Sato et al. 2013, 3). This problem was worst in sub-Saharan Africa, but also in parts of Latin America, Asia and Europe.

The treatment of faeces and urine (as opposed to other types of wastewater) can be broken into two categories: decentralised and centralised. Decentralised 'on-site' approaches are more common in lower-income countries and rural settings and city 'slums' that lack sewerage infrastructure (WWAP 2017). In broad terms, decentralised treatment is an indicator of community poverty – an important point we return to below. Examples of decentralised systems include open defecation, dry pits, septic tanks and community-based treatments plants that serve up to 100 households (Kerstens et al. 2016). Centralised 'off-site' approaches include community-based treatment systems that service up to 5,000 households as well as large-scale sewerage networks serving 50,000 households (Kerstens et al. 2016) or more.

Centralised systems operate in all regions of the world, although the degree to which countries effectively *treat* wastewater varies (e.g. Sato et al. 2013). Peal et al.'s (2014) tentative findings from a study of 12 cities in lower-income countries showed considerable variation in the percentage of households connected to a centralised system. Because of the difficulties in accessing data from government agencies in each city, Peal et al. (2014) based their estimates on secondary data and qualitative information. Despite the authors' caveats about their figures, the city-level comparison is useful. On the one hand, it shows that WWA research can be very challenging, or impossible, in some cities. For instance, Palu in Indonesia does not appear to have any sewerage infrastructure. A similar situation exists in Dumaguete in the Philippines.

Another complexity is apparent from Peal et al.'s (2014) study. Since decentralised systems (on-site and open defecation) are more common in poorer communities, WWA researchers in some cities would need to exercise caution to avoid *selection effects*. For example, a WWA study from Manila's centralised system may produce data that predominantly reflects wealthier Filipinos' consumption of drugs, alcohol and tobacco. In other words, the data might not be representative of the general population. This is not to discount the possibility that centralised systems in some lower-income cities do service communities with different socio-economic profiles. But failing to account for selection effects would undermine the value of WWA studies in lower-income counties.

On the other hand, functioning sewerage infrastructure clearly does exist in the cities of many lower-income countries (Peal et al. 2014). The decision on the viability of WWA research in any of these cities would need to be based on site inspections and testing by sewerage engineers. Population size would also be a relevant consideration. In broad terms, all things being equal, catchments that serve large populations are more valuable for monitoring purposes than smaller ones because they are more representative of the general population and provide more stable estimates. Notably, Delhi has a population of many millions of people and approximately 75% of households are connected to the sewers (Peal et al. 2014). This suggests that in Delhi researchers would have good prospects of finding catchments in which WWA sampling was feasible and so could produce data on the consumption of tobacco, alcohol and other drugs by a large numbers of people.

Other cities in which fewer households are connected to sewerage may still be able to accommodate useful WWA research because of their population sizes. For instance, 25% of households in Dakar (Senegal) are connected to sewerage. In 2012 terms, this represents over 600,000 people in the city's population of about 2.47 million (Peal et al. 2014).

Nairobi, Kenya's capital, was not included in Peal et al.'s (2014) research. Nairobi is noteworthy because it is serviced by the Dandora Sewage Treatment Plant. This plant is reportedly the largest wastewater stabilisation pond in Africa.[8] It is estimated to service 80% of Nairobi's population – or approximately 2.5 million people.[9] Wastewater analysis conducted at this plant could be of considerable value for global research purposes because of the paucity of data on sub-Saharan African consumption of illicit drugs, alcohol and tobacco. Naturally, these data would be beneficial for policymakers within Kenya as a lower-middle–income country.

Another good example of a WWTP that could house useful WWA research is the As-Samra Wastewater Treatment Plant in Jordan, which is a lower-middle–income country (World Bank 2017). This plant services a catchment of around 2.27 million people (UNESCWA 2015).

Finally, India is also classified as a lower-middle–income country (World Bank 2017). With a population approaching 1.4 billion people, it is not surprising that India's sewerage infrastructure dwarfs that of many high-income countries. India's most recent inventory of sewage treatment plants found that 522 are operational in cities across the nation (CPCB 2015). This suggests good possibilities for WWA at multiple sites in that country.

3.4.1.2 Complexities Relating to Sewerage Performance, Climate and Substance Use Practices

As the number of countries conducting WWA has expanded, new technical complexities have been encountered. This reflects the fact that many different systems are employed to collect, treat and discharge wastewater (Tilley et al. 2014). Devault et al. (2014, 2017) conducted the first WWA research in the Caribbean in Fort de France, the capital of Martinique – a small island and French territory. Martinique is not a country so the World Bank has not classified its income status. But several aspects of the studies by Devault and colleagues reflect the sorts of conditions that WWA researchers may encounter in lower-income countries that differentiate it from studies in upper-middle income countries such as Brazil (Maldaner et al. 2012), Colombia (Bijlsma et al. 2016) and South Africa (Archer et al. 2018).

The population of Fort de France is approximately 110,000 people. The team reported that the city's sewer system serviced a total of 50,000 inhabitants. Samples were collected from this sewer network at four WWTPs servicing approximately 47,200 people. Official figures on the population of each of the four catchments were not available. Consequently the study relied upon a method for estimating population based on hydro-chemical parameters.

The four WWTPs respectively treated wastewater from: a middle-class district; two "poor" populations; and part of a slum district (Devault et al. 2014, 972). While the socio-economic profiles of these catchments were diverse, it is doubtful that the samples collected by Devault et al. (2014) were representative of the population of Fort de France. This is because households not connected to the sewer system tend to be poorer.

In high-income countries, water authorities can provide information that is important for WWA calculations, such as the rate at which sewers leak and the extent to which the sewers collect rainwater (see further 2.2). This detail did not appear to be available for Devault et al. (2014). In part this was due to (a) sewer-damage due to seismic activity, (b) the comparatively rapid degradation of sewer pipes in the local climate, and (c) illegal connections that channelled rainwater into the network (Devault et al. 2014; Devault, Lévi and Karolak 2017).

Two other important issues were identified by Devault et al. (2014). Because of the tropical climate and the relatively high temperature of wastewater (26–31 °C, see Devault, Lévi and Karolak 2017), the researchers needed to account for the fact that biomarkers degrade faster than in colder climates. More complicated factors related to the type of cocaine used in Martinique – crack cocaine – and the way it is consumed. At the time of the study little was known about the way the human body metabolises crack cocaine. Local practices for smoking crack cocaine may influence metabolic functioning. These included: piping crack cocaine through rum instead of water; and smoking crack cocaine together with cannabis ("black joint form") (Devault et al. 2014, 976). Cone et al. (1998) and Castiglioni et al. (2013) reported that the main biomarker of cocaine usage (benzoylecgonine) is more than twice as abundant in the urine of intravenous users than in those who smoke it. Crack or free-base cocaine is smoked rather than injected. Clearly it is desirable that the back calculation of cocaine consumption from sewage benzoylecgonine loads takes account of the ratio of cocaine intravenous users to smokers in a community. This approach was used by Devault et al. (2014).

Further research has been conducted by Devault and colleagues (2017) in Martinique – to estimate drug consumption and assess the degree to which local WWTPs successfully prevent pollutants from entering the environment. The team is refining its methods for calculating the fate of biomarkers in conditions typical of tropical wastewater. They estimated that "sewer losses reach 50% of the volume in the studied sewage" (Devault, Lévi and Karolak 2017, 253).

3.4.2 *WWA in Lower-Income Countries in the Short, Medium and Long Term*

Devault et al. (2014, 2017) give a dispassionate account of the complexities WWA researchers may encounter in lower-income countries. This team is to be lauded for its sustained effort to refine WWA in these challenging conditions. What are the implications of their experience for global comparative analyses? There are several points we want to make.

First, despite the caveats attached to the Martinique studies, Devault and colleagues have produced quantitative data in a location that has previously received very little empirical attention. Broadly, as reported by the UNODC (2017b), the findings indicate that cocaine consumption on the island is comparable to WWA estimates in Medellin, Colombia (Bijlsma et al. 2016) and higher than estimates from European WWA data (Devault et al. 2014). The WWA estimates from Martinique may change over time as the research team refines its approach. But the significance of the opportunity to build trend data in a previously under-studied location should not be understated.

The Martinique studies also demonstrate that WWA projects can be managed over long distances. Water authorities assisted with collecting samples in Fort de France (Devault et al. 2014). These were frozen and transported 6,700 km to France for analysis. Similar approaches have been adopted elsewhere (e.g. Archer et al. 2018). This allows lower-income countries to use WWA without local experts in analytical chemistry or high-tech laboratories.

With so little WWA in lower-income countries it is unclear whether the conditions encountered in Martinique are typical or atypical. The information discussed above in 3.4.1 suggested that many lower-middle–income countries operate WWTPs that could be very suitable for WWA – both in terms of technical feasibility and in terms of the value of the WWA data they could yield for global monitoring of drugs, alcohol and tobacco. Over 500 WWTPs operate in India alone. The Dandora plant in Nairobi services about 2.5 million Kenyans. If the plant was suitable for WWA, it could provide rich annual data about psychoactive substances consumption in sub-Saharan Africa – the region of the world in which the United Nations and other agencies encounter greatest difficulty in data gathering.

Importantly, there are good reasons why agencies with carriage of WWTPs in lower-income countries might be interested in facilitating WWA research. This is because WWA – which originated from the larger field of environmental toxicology – can also be used to measure the performance of WWTPs in their capacity to extract pollutants from wastewater before releasing it into the environment. In our experience in Australia, the capacity of WWA to measure the environmental performance of WWTPs is a key reason why water agencies support the national WWA monitoring program.

Finally, in all likelihood the numbers of WWA-viable sites will increase in coming years, perhaps markedly in some regions. For example, a relatively recent inventory of India's sewerage infrastructure showed that 145 WWTPs were under construction and a further 70 were proposed for construction (CPCB 2015). Modelling on the situation in Indonesia suggested that by 2035 one-third of the community could be connected to a medium-centralised or centralised system (Kerstens et al. 2016). Six years ago, the World Bank (2012; cited in WWAP 2017, 88) estimated that "half of the urban infrastructure that will make up African cities by 2035 has yet to be built".

3.5 Conclusion

This chapter traversed a broad range of topics relating to illicit drugs, tobacco and alcohol. In discussing these three classes of substances together we have departed from the separation of these substances in the academic literature and international policy (e.g. UNODC 2017b). The chapter overviewed global metrics on the scale of the markets for illicit drugs, tobacco and alcohol, and examined problems in deriving these metrics using traditional methods. It highlighted that key global policy objectives are assessed by comparisons of countries' descriptive data because quasi-experimental designs are not feasible at the international level.

The chief aim of this chapter was to explore whether WWA can assist the international community to monitor consumption of psychoactive substances. Immense sums of money are spent globally on wastewater infrastructure each year. Heymann, Lizio and Siehlow (2010) estimated the annual figure to be US$104 billion. The United States Environmental Protection Agency (2016) has estimated that US$271 billion is required to meet the country's wastewater infrastructure needs, including refurbishing existing systems. While monetary values were not available for India, the investment that country is making in wastewater infrastructure is enormous; as noted, 145 WWTPs are under construction and a further 70 are proposed.

It is prudent and rational to investigate whether WWA can capitalise on this investment. In many respects the methodological problems with traditional methods of monitoring national substance use are exacerbated at the international level. In addition to the difficulties in comparing data derived from within countries, the production of data for global monitoring is slow. Crucially, data are missing or are inadequate in many regions of the world whose countries are least able to bear the economic and social burdens of psychoactive substance use. SCORE and the EMCDDA have illustrated that international comparison with WWA data has moved beyond proof of concept. The use of WWA data by the UNODC indicates that policymakers and researchers outside of the WWA field have accepted the efficacy of the method as an additional and useful tool for drug monitoring.

Centralised sewerage infrastructure does not exist or is limited in many lower-income countries. The Martinique studies show that the selection of new sites for WWA ought to involve careful site inspections by sewer engineers. It still seems likely that WWA is feasible in multiple WWTPs in lower-income countries and we used examples from Kenya, Jordan and India to support this claim. There are good reasons to believe that possibilities for WWA will increase as more WWTPs are built in the next 20 years or so.

Consistent with the perspective we presented in Chapter 2, we have not argued that WWA can replace traditional methods. The efforts of agencies such as the UNODC and WHO to build capacity in traditional research methods in lower-income countries need to be pursued. However, it is likely that far fewer resources will be required to establish protocols for WWA sampling in lower-income countries. Negotiations with relevant authorities may be simplified because WWA researchers can also provide information about WWTPs' core business with respect to treating sewage water.

The implication of this chapter for WWA scholars is that significant global benefit could be gained by extending their research into lower-income countries. In our view this will place even greater value on the endeavours of WWA teams to develop their own scientific methods to estimate population size (see 2.2.4) – primarily because reliable demographic statistics may not exist in lower-income settings. Other areas of scientific research have been underscored in this chapter too. Chief among these is the importance of refining methods for estimating the consumption of alcohol and tobacco. More needs to be learned about the degradation of biomarkers in various climates and sewer designs. Local practices around substance use – routes of administration, poly-drug use and so forth – may present novel challenges in understanding metabolic functioning.

Notes

1. *Single Convention on Narcotic Drugs* (1961); *Convention on Psychotropic Substances* (1971); *Convention against Illicit Traffic in Narcotic Drugs and Psychotropic Substances* (1988).

2. An explanation of the steps involved in estimating heroin production rates is provided in table notes on page xi of the Annex in UNODC (2016a).
3. See table notes provided on page vi of the Annex.
4. See World Health Organization. Global Information System on Alcohol and Health (GISAH). www.who.int/gho/alcohol/en/.
5. See International Tobacco Control (ITC) Policy Evaluation Project. Accessed 19 April 2018. www.itcproject.org/. Note also the WHO role in the Global Tobacco Surveillance System: Centers for Disease Control and Prevention. About GTSS. Accessed 19 April 2018. www.cdc.gov/tobacco/global/gtss/index.htm. For a complete list of national surveys accessed by the WHO see WHO (2017b, 211–35).
6. World estimates suggest that about 346 million people use smokeless tobacco and most of these live in Southeast Asia (USNCI and WHO 2016).
7. Lubertus Bijlsma (University Jaume I, Spain) and Barbara Kasprzyk-Hordern (University of Bath, UK).
8. See SMEC. Dandora Sewage Treatment Plant. Accessed 14 May 2018. www.smec.com/what-we-do/projects/Dandora-Sewage-Treatment-Plant.
9. Nairobi's population is approximately 3.13 million people. See United Nations Statistics Division (2016).

References

Archer, E., E. Castrigano̒, B. Kasprzyk-Hordern, and G. M. Wolfaardt. 2018. Wastewater-based epidemiology and enantiomeric profiling for drugs of abuse in South African wastewater. *Science of the Total Environment* 625:792–800.

Australian Criminal Intelligence Commission. 2019. National Wastewater Drug Monitoring Program – Report 7. www.acic.gov.au/sites/default/files/2019/06/nwdmp7_140619.pdf?v=1560498324.

Australian Criminal Intelligence Commission. 2020. National Wastewater Drug Monitoring Program – Report 9. www.acic.gov.au/publications/reports/national-wastewater-drug-monitoring-program-ninth-report

Babor, T. F., R. Caetano, S. Casswell et al. 2010a. *Alcohol: No ordinary commodity: Research and public policy*. Oxford: Oxford University Press.

Babor, T. F., J. P. Caulkins, G. Edwards et al. 2010b. *Drug policy and the public good*. Oxford: Oxford University Press.

Baker, D. R., V. Očenášková, M. Kvicalova, and B. Kasprzyk-Hordern. 2012. Drugs of abuse in wastewater and suspended particulate matter – Further developments in sewage epidemiology. *Environment International* 48:28–38.

Banta-Green, C. J., J. A. Field, A. C. Chiaia, D. L. Sudakin, L. Power, and L. de Montigny. 2009. The spatial epidemiology of cocaine, methamphetamine and 3,4-methylenedioxymethamphetamine (MDMA) use: A demonstration using a population measure of community drug load derived from municipal wastewater. *Addiction* 104:1874–80.

Bijlsma, L., A. M. Botero-Coy, R. J. Rincón, G. A. Peñuelac, and F. Hernándeza. 2016. Estimation of illicit drug use in the main cities of Colombia by means of urban wastewater analysis. *Science of The Total Environment* 565:984–93.

Bones, J., K. V. Thomas, and B. Paull. 2007. Using environmental analytical data to estimate levels of community consumption of illicit drugs and abused pharmaceuticals. *Journal of Environmental Monitoring* 9:701–7.

Burgard D. A., C. Banta-Green, and J. A. Field. 2014. Working upstream: How far can you go with sewage-based drug epidemiology?. *Environmental Science and Technology* 48:1362–68.

Burris, S. C. 2017 Theory and methods in comparative drug and alcohol policy research: Response to a review of the literature. *International Journal of Drug Policy* 41:126–31.

Castiglioni, S., L. Bijlsma, A. Covaci et al. 2013. Evaluation of uncertainties associated with the determination of community drug use through the measurement of sewage drug biomarkers. *Environmental Science & Technology* 47:1452–60.

Central Pollution Control Board. 2015. Inventorization of sewage treatment plants. https://nrcd.nic.in/writereaddata/FileUpload/NewItem_210_Inventorization_of_Sewage-Treatment_Plant.pdf.

Cone, E. J., A. Tsadik, J. Oyler, and W. D. Darwin. 1998. Cocaine metabolism and urinary excretion after different routes of administration. *Therapeutic Drug Monitoring* 20(5):556–60.

Decorte, T., D. Mortelmans, J. Tieberghien, and S. De Moor. 2009. *Drug use: An overview of general population surveys in Europe*. Luxembourg: Office for Official Publications of the European Communities.

Degenhardt, L., L. Dierker, W. T. Chiu et al. 2010. Evaluating the drug use 'gateway' theory using cross-national data: Consistency and associations of the order of initiation of drug use among participants in the WHO World Mental Health Surveys. *Drug and Alcohol Dependence* 108:84–97.

Devault, D. A., Y. Lévi, and S. Karolak. 2017. Applying sewage epidemiology approach to estimate illicit drug consumption in a tropical context: Bias related to sewage temperature and pH. *Science of the Total Environment* 584–585:252–58.

Devault, D. A., T. Néfau, H. Pascaline, S. Karolak, and Y. Lévi. 2014. First evaluation of illicit and licit drug consumption based on wastewater analysis in Fort de France urban area (Martinique, Caribbean), a transit area for drug smuggling. *Science of The Total Environment* 490:970–78.

Doll, R., R. Peto, J. Boreham, and I. Sutherland. 2004. Mortality in relation to smoking: 50 years' observations on male British doctors. *British Medical Journal* 328:1519–27.

EMCDDA. 2018. *Perspectives on drugs: Wastewater analysis and drugs: A European multi-city study*, updated 7 March 2018. Lisbon: EMCDDA.

Enzmann, D., J. Kivivuori, I. H. Marshall, M. Steketee, M. Hough, and M. Killias. 2018. Introduction to the International Self-Report Delinquency Study (ISRD3). In *A global perspective on young people as offenders and victims*, 1–6. Cham: Springer.

European Monitoring Centre for Drugs and Drug Addiction (EMCDDA). 2016. *Assessing illicit drugs in wastewater: Advances in wastewater-based drug epidemiology*. Insights 22. Luxembourg: Publications Office of the European Union.

Evans-Lacko, S., S. Aguilar-Gaxiola, A. Al-Hamzawi et al. 2018. Socio-economic variations in the mental health treatment gap for people with anxiety, mood, and substance use disorders: Results from the WHO World Mental Health (WMH) surveys. *Psychological Medicine* 48:1560–71.

Giommoni, L., P. Reuter, and B. Kilmer. 2017. Exploring the perils of cross-national comparisons of drug prevalence: The effect of survey modality. *Drug and Alcohol Dependence* 181:194–99.

Glynn, M. K., and L. C. Backer. 2010. Collecting public health surveillance data: Creating a surveillance system. In *Principles and practice of public health surveillance* 3rd edition, ed. L. M. Lee, S. M. Teutsch, S. B. Thacker, and M. E. St. Louis, 44–64, New York: Oxford University Press.

Gonzalez-Marino, I., J.A. Baz-Lomba, , N.A. Alygizakis et al. 2019. Spatio-temporal assessment of illicit drug use at large scale: Evidence from 7 years of international wastewater monitoring. *Addiction* 115:109–20.

Hall, W. 2018. The future of the international drug control system and national drug prohibitions. *Addiction* 113:1210–23.

Harrison, L., and A. Hughes. 1997. *The validity of self-reported drug use: Improving the accuracy of survey estimates*. Rockville, MD: U.S. Department of Health and Human Services, National Institutes of Health, National Institute on Drug Abuse, Division of Epidemiology and Prevention Research.

Heymann, E., D. Lizio, and M. Siehlow. 2010. *World water markets: High investment requirements mixed with institutional risk*. Deutsche Bank Research. www.dbresearch.com/PROD/RPS_EN-PROD/PROD0000000000474405/World_water_markets%3A_High_investment_requirements_.pdf.

Johnson, T. P. 2015. Measuring substance use and misuse via survey research: Unfinished business. *Substance Use & Misuse* 50:1134–38.

Kerstens, S. M., M. Spiller, I. Leusbrock, and G. Zeeman. 2016. A new approach to nationwide sanitation planning for developing countries: Case study of Indonesia. *Science of the Total Environment* 550:676–89.

Khan, U., A. L. N. van Nuijs, J. Li et al. 2014. Application of a sewage-based approach to assess the use of ten illicit drugs in four Chinese megacities. *Science of the Total Environment* 487:710–21.

Kilmer, B., P. Reuter, and L. Giommoni. 2015. What can be learned from cross-national comparisons of data on illegal drugs?. *Crime and Justice* 44:227–96.

Kim, K. Y., F. Y. Lai, H. Kim, P. K. Thai, J. F. Mueller, and J. Oh. 2015. The first application of wastewater-based drug epidemiology in five South Korean cities. *Science of the Total Environment* 524:440–46.

Kraus, L., U. Guttormsson, H. Leifman et al. 2016. *ESPAD Report 2015: Results from the European School Survey Project on alcohol and other drugs*. Luxembourg: Publications Office of the European Union.

Lai, F. Y., C. Gartner, W. Hall et al. 2018. Measuring spatial and temporal trends of nicotine and alcohol consumption in Australia using wastewater-based epidemiology. *Addiction* 113:1127–36.

Lai, F. Y., P. K. Thai, J. O'Brien et al. 2013. Using quantitative wastewater analysis to measure daily usage of conventional and emerging illicit drugs at an annual music festival. *Drug and Alcohol Review* 32:594–602.

Maldaner, A. O., L. L. Schmidt, M. A. F. Locatelli et al. 2012. Estimating cocaine consumption in the Brazilian Federal District (FD) by sewage analysis. *Journal of the Brazilian Chemical Society* 23:861–67.

Metcalfe, C., K. Tindale, H. Li, A. Rodayan, and V. Yargeau. 2010. Illicit drugs in Canadian municipal wastewater and estimates of community drug use. *Environmental Pollution* 158:3179–85.

Moriame E., and A. Greliche. 2007. Les comptes économiques de la Martinique en 2006. Accessed 1 May 2018. www.epsilon.insee.fr/jspui/bitstream/1/15720/1/cerom_mart_comptes_2007.pdf

Nefau, T., S. Karolak, L. Castillo, V. Boireau and Y. Levi. 2013. Presence of illicit drugs and metabolites in influents and effluents of 25 sewage water treatment plants and map of drug consumption in France. *Science of the Total Environment* 461:712–22.

Östman, M., J. Fick, E. Nässtróm, and R. H. Lindberg. 2014. A snapshot of illicit drug use in Sweden acquired through sewage water analysis. *Science of the Total Environment* 472:862–71.

Peal, A., B. Evans, I. Blackett, P. Hawkins, and C. Heyman. 2014. Fecal sludge management: A comparative assessment of 12 cities. *Journal of Water, Sanitation and Hygiene for Development* 4:563–75.

Rehm, J., S. Kailasapillai, E. Larsen et al. 2014. A systematic review of the epidemiology of unrecorded alcohol consumption and the chemical composition of unrecorded alcohol. *Addiction* 109: 880–93.

Rehm, J., F. Kanteres, and D. W. Lachenmeier. 2010. Unrecorded consumption, quality of alcohol and health consequences. *Drug and Alcohol Review* 29:426–36.

Ritter, A., M. Livingston, J. Chalmers, L. Berends, and P. Reuter. 2016. Comparative policy analysis for alcohol and drugs: Current state of the field. *International Journal of Drug Policy* 31:39–50.

Saha, S., J. G. Scott, D. Varghese, L. Degenhardt, T. Slade, and J. J. McGrath. 2011. The association between delusional-like experiences, and tobacco, alcohol or cannabis use: A nationwide population-based survey. *BMC Psychiatry* 11:202.

Sato, T., M. Qadir, S. Yamamoto, T. Endo, and A. Zahoor. 2013. Global, regional, and country level need for data on wastewater generation, treatment, and use. *Agricultural Water Management* 130:1–13.

Singleton, N., A. Cunningham, T. Groshkova, L. Royuela, and R. Sedefov. 2018. Drug supply indicators: Pitfalls and possibilities for improvements to assist comparative analysis. *International Journal of Drug Policy* 56:131–36.

Substance Abuse and Mental Health Services Administration. 2017. *Key substance use and mental health indicators in the United States: Results from the 2016 National Survey on Drug Use and Health* (HHS Publication No. SMA 17–5044, NSDUH Series H-52). Center for Behavioral Health Statistics and Quality, Substance Abuse and Mental Health Services Administration. www.samhsa.gov/data/sites/default/files/NSDUH-FFR1-2016/NSDUH-FFR1-2016.htm.

Terzic, S., I. Senta, and M. Ahel. 2010. Illicit drugs in wastewater of the city of Zagreb (Croatia) – Estimation of drug abuse in a transition country. *Environmental Pollution* 158: 2686–93.

Tilley, E., L. Ulrich, C. Lüthi, P. Reymond, and C. Zurbrügg. 2014. *Compendium of sanitation systems and technologies* 2nd revised edition. Dübendorf: Swiss Federal Institute of Aquatic Science and Technology (Eawag).

United Nations. 2015. *Transforming our world: The 2030 Agenda for Sustainable Development*. https://sustainabledevelopment.un.org/content/documents/21252030%20Agenda%20for%20Sustainable%20Development%20web.pdf.

United Nations Children's Fund, and World Health Organization. 2015. *Progress on sanitation and drinking water: 2015 update and MDG assessment*. New York: United Nations Children's Fund; Geneva: World Health Organization.

United Nations Convention against Illicit Traffic in Narcotic Drugs and Psychotropic Substances. 1988. www.unodc.org/unodc/en/treaties/illicit-trafficking.html

United Nations Convention on Psychotropic Substances. 1971. www.unodc.org/unodc/en/treaties/psychotropics.html

United Nations Economic and Social Commission for Western Asia. 2015. *ESCWA water development report 6: The water, energy, food security nexus in the Arab region*. Beirut: United Nations.

United Nations Office on Drugs and Crime. 2015. *World drug report 2015*. New York: United Nations.

United Nations Office on Drugs and Crime. 2016a. *World drug report 2016*. New York: United Nations.

United Nations Office on Drugs and Crime. 2016b. UNODC annual report: Covering activities during 2016. www.unodc.org/documents/AnnualReport2016/2016_UNODC_Annual_Report.pdf.

United Nations Office on Drugs and Crime. 2017a. *World drug report 2017 (Booklet 1 – Executive summary: Conclusions and policy implications)*. New York: United Nations.

United Nations Office on Drugs and Crime. 2017b. *World drug report 2017 (Booklet 4 – Market analysis of synthetic drugs: Amphetamine-type stimulants, new psychoactive substances)*. New York: United Nations.

United Nations Office on Drugs and Crime. 2018. *World drug report 2018 (Booklet 3 – Analysis of drug markets: Opiates, cocaine, cannabis, synthetic drugs)*. New York: United Nations.

United Nations Office on Drugs and Crime. 2019. *World drug report 2019 (Booklet 4 – Stimulants)*. New York: United Nations.

United Nations Single Convention on Narcotic Drugs. 1961. www.unodc.org/unodc/en/treaties/single-convention.html

United Nations Statistics Division. 2016. City population by sex, city and city type. Accessed 10 March 2020. http://data.un.org/Data.aspx?q=kenya+population&d=POP&f=tableCode%3A240%3BcountryCode%3A404.

United States Environmental Protection Agency. 2016. Clean Watersheds Needs Survey 2012: Report to Congress. www.epa.gov/sites/production/files/2015-12/documents/cwns_2012_report_to_congress-508-opt.pdf.

United States National Cancer Institute, and World Health Organization. 2016. *The Economics of Tobacco and Tobacco Control*. Tobacco Control Monograph 21. National Cancer Institute http://cancercontrol.cancer.gov/brp/tcrb/monographs/21/index.html.

Wilkins, C., F.Y. Lai, J. O'Brien, P. Thai and J.F. Mueller. 2018. Comparing methamphetamine, MDMA, cocaine, codeine and methadone use between the Auckland region and four Australian states using wastewater-based epidemiology (WBE). *New Zealand Medical Journal* 131(1478):12–20.

World Bank. 2017. World Bank Country and Lending Groups (2017 listing). Accessed 1 May 2018. https://datahelpdesk.worldbank.org/knowledgebase/articles/906519-world-bank-country-and-lending-groups.

World Health Organization. 2003. WHO Framework Convention on Tobacco Control. www.who.int/fctc/text_download/en/.

World Health Organization. 2006. *Guidelines for the safe use of wastewater, excreta and greywater: Volume 3 – Wastewater and excreta use in aquaculture*. Geneva: World Health Organization.

World Health Organization. 2010. *Global strategy to reduce the harmful use of alcohol*. Geneva: World Health Organization.

World Health Organization. 2014. *Global status report on alcohol and health 2014*. Geneva: World Health Organization.

World Health Organization. 2017a. *World health statistics 2017: Monitoring health for the SDGs, Sustainable Development Goals*. Geneva: World Health Organization.

World Health Organization. 2017b. *WHO report on the global tobacco epidemic 2017: Monitoring tobacco use and prevention policies*. Geneva: World Health Organization.

WWAP (United Nations World Water Assessment Program). 2017. *The United Nations World Water Development Report 2017 – Wastewater: The Untapped Resource*. Paris: United Nations Educational, Scientific and Cultural Organization.

chapter four

Meso Applications of Wastewater Analysis

National Research

INTRODUCTION

It is worth reiterating here that, as stated in Chapter 1, WWA can be used to examine a wide range of issues relating to population health and the environment. However, the focus of this book is on WWA in the context of studying substance use.

The focus of Chapter 3 was the potential utility of wastewater analysis (WWA) *globally*. On the international stage the most practical form of evaluative research is cross-country comparison of descriptive statistics derived from surveys and event data in combination with more qualitative information such as expert opinion. We examined how WWA is being used to compare countries' estimated consumption of illicit drugs by the European Monitoring Centre for Drugs and Drug Addiction (EMCDDA 2019) and, more latterly, the United Nations Office on Drugs and Crime (UNODC 2017b). The thesis in Chapter 3 was that WWA could be considerably useful in lower-income countries in monitoring not only illicit drugs but also alcohol and tobacco.

Chapter 4 moves the discourse from the international to the national level. We examine the utility of WWA in countries that have established sewerage infrastructure. In our view, there is a strong case for WWA providing an additional tool for monitoring the consumption of psychoactive substances within countries (Castiglioni and Vandam 2016) based on research showing that the approach has worked effectively in Europe and Australia.

What other opportunities does WWA offer? To get a sense of this, it is first important to note the very large scale of sewer networks in some countries. As seen in Chapter 3, India was recently estimated to have 522 operational wastewater treatment plants (WWTPs) (CPCB 2015). In the USA approximately 3,177 publicly owned WWTPs service a population of 10,000 people or more (US EPA 2016a). (As explained later, a population of 10,000 represents a threshold above which WWA estimates have acceptable uncertainty (Castiglioni and Lor 2016, 36).) A further 11,571 smaller publicly owned WWTPs operate in America (US EPA 2016a). Together, all 14,748 WWTPs service 238.2 million people (EPA 2016a), which is 73% of the American population. In fact, Texas alone operates more WWTPs than India: 551 (US EPA 2016b). American households are connected to treatment plants by approximately 1.2 million kilometres of public sewers (Sterling et al. 2010).

Similar levels of investment in sewerage infrastructure can be found elsewhere. Close to 25 million properties in England and Wales are serviced by 565,070 km of sewers and 6,354 WWTPs (Discover Water UK 2018). Best estimates suggest that Australia has 673 urban WWTPs – each of which services over 10,000 connections (BOM 2017; cited in BITRE 2017, 289–290).

For wastewater researchers – scientists and social scientists alike – these statistics have implications for the management of research projects. Many WWTPs may be a viable sampling point for WWA. Sewer pipes represent kilometres of data collection apparatus.

For policy makers, these statistics underscore the enormity of the investment already made in sewer infrastructure, to say nothing of recurrent expenditure on its maintenance and upgrades. In a report to the US Congress, the Environmental Protection Agency (EPA) (2016a) estimated that $271 billion was required over a 20-year period for capital investment in infrastructure to manage wastewater and stormwater. Wastewater analysis offers a completely different way to obtain additional benefit for the community from the significant investment of public funds in sewer infrastructure.

In this chapter we argue that WWA can be of substantial value in two main ways for 'domestic research' on drug use, i.e. research within countries. The first relates to monitoring substance use in rural settings. After discussing the difficulties in defining what 'rural' is we (a) examine why rural substance use is an important topic in its own right that is distinct from drug use in urban contexts; (b) explain the practical limitations of traditional survey methods for monitoring drug use in rural areas; and (c) argue that WWA can overcome some of these limitations when there are suitable sites for sampling.

We explore the concept of the 'data shadow' that affects many rural communities (Prichard et al. 2018, 197). This term refers to the fact that many rural areas fall outside monitoring systems and are infrequently studied by government and academic researchers. Much of the research that is analysed in this chapter originated from epidemiology. But we also draw on research in rural criminology.

America and Australia feature in this chapter. This makes our analyses manageable by restricting their scope while considering two countries that are useful to contrast. Although America's population is over 13 times that of Australia, both countries are very large geographically. They are the third and sixth largest countries by size in the world, respectively. Additionally, both America and Australia have world-class systems for monitoring substance use using traditional survey methods. However, unlike America, Australia has established a nationwide WWA monitoring system, which includes some rural sampling sites. Finally, in both countries concerns have been raised about the impact of substance use in rural populations, specifically methamphetamines and opioids.

The second benefit we discuss is the possibility of conducting quasi-experimental evaluations of policies using WWA (see 1.3). Granularity of data is key: because WWA data can provide a far greater number of data points than traditional survey methods, it provides greater statistical power for quasi-experimental research. There are good reasons to believe that quasi-experimental WWA research in rural settings could be *empirically superior* to that in urban settings (Prichard et al. 2018). In short, it is argued that whereas rural isolation is a disadvantage with the use of traditional survey methods, it may well be an advantage for WWA research in some rural communities.

4.1 Substance Use in Rural Settings

In 2016 the Drug Prevention and Health Branch of the UNODC and the World Health Organization (WHO) convened an international workshop on rural substance use (Milano et al. 2017). The workshop involved participants from 20 countries[1] whose collective experience was that there is very little evidence available on substance use in rural settings.

The participants noted that the term 'rural' is difficult to define. Ultimately perceptions of what is 'rural' may differ markedly between cultures and can be influenced by geographic location, population density, the prevalence of identifiable rural subcultures, the dominance of agriculture as a source of employment and so forth (Milano et al. 2017).

Within countries, the appropriate definition of rural may differ according to policy objectives or the focus of research projects. For instance the Economic Research Service of the United States Department of Agriculture has developed a variety of systems for classifying a region as rural at the county and sub-county level (see USDA ERS 2018a). At the county level one classification is the Natural Amenities Scale. This is "a measure of the physical characteristics of a county" that incorporates "climate, topography, and water area that reflect environmental qualities most people prefer" (USDA ERS 2018b).

Another classification system is the Rural-Urban Continuum Code, which splits counties into metropolitan areas (with three sub-categories based on population size) and nonmetropolitan areas (with six sub-categories that reflect population size and proximity to a metro area). This code is used in substance use monitoring – notably the annual American National Survey on Drug Use and Health (NSDUH) (SAMHSA 2017a).

4.1.1 *Challenges for Rural Research in Criminology and Epidemiology*

We know that aspects of urban environments are associated with crime – just two examples are overcrowding and higher costs of living (see Spooner and Hetherington 2004). However, criminologists have criticised their own discipline for ignoring rural settings (Barclay and Donnermeyer 2002). Multiple factors explain criminology's urban-centric bias in studying drug markets and other forms of crime. Carrington, Donnermeyer and DeKeserdy (2014) have argued that influential American writing from the early 20th century promoted the view that cities are more criminogenic than non-urban environments. This perspective ultimately "privileged the city as the ideal research laboratory" (Carrington, Donnermeyer and DeKeserdy 2014, 464). By contrast there was an idealised perspective in which rural places benefitted from healthy lifestyles and high social capital marked by "qualities of friendliness, togetherness, honesty and low crime" (Barclay, Donnermeyer and Jobes 2004, 7).

Lack of resources is another reason for the stunted growth of rural criminological research. Rural research is often more difficult than that in urban areas. To begin with, criminologists based in universities or government research centres tend to work in cities, making the study of urban crime less time-consuming than going further afield. In geographically large countries with dispersed rural populations, research in rural areas can be expensive and slow to conduct (Meit et al. 2014, 5; Prichard et al. 2018). Recruiting participants can be particularly challenging given locals' reluctance to talk "to outsiders about drug issues" (Delahunty and Putt 2006, 3). In fact, in interviews with 620 farmers about agricultural crime, Barclay, Donnermeyer and Jobes (2004) concluded that interconnectedness in rural communities can (a) promote tolerance for certain types of crime and (b) generate strong social taboos about reporting aspects of crime to authorities.

Health researchers also accept that rural issues have been understudied, particularly before the 1990s (Smith, Humphreys and Wilson 2008). Where drug use is concerned, this may reflect the belief that rural communities are not at risk of problematic drug use (Gfroerer, Larson and Colliver 2007; Milano et al. 2017). Structural problems have been examined too, in terms of the lack of funding for rural research. Analyses of the studies funded by Australia's National Health and Medical Research Council suggest that resourcing for rural research does not match the size of the populations living in rural areas (Barclay, Phillips and Lyle 2018).

Practical challenges have also been identified in conducting health research in rural areas. These include: distance, locating participants and overcoming the perception of city researchers as outsiders (e.g. Jan and Khan 2015; Pierce and Scherra 2004; Wilkes 1999; Spooner, Bishop and Parr 1997).

In its monograph, *The 2014 Update of the Rural-Urban Chartbook,* the American Rural Health Reform Policy Research Center drew on multiple data sources on national health. Because of "the high cost of collecting data in sparsely populated areas" the Center found that many health surveys incorporated a small number of rural counties (Meit et al. 2014, 5). In some instances, reliable estimates could not be made even when the data were combined from all rural counties. Further complexities related to missing data (e.g. county of residence) and the lack of data granularity at the sub-county level (Meit et al. 2014; see also Gfroerer, Larson and Colliver 2007).

4.1.2 *Features of Rural Substance Use*

In geographically large countries it is impossible for governments to provide services and amenities to the same standard in all places. This can have significant implications for everyday life. As rural settings become more isolated and sparsely populated, inequalities in the administration of the criminal justice system can increase (Hogg 2011). Some rural communities can be disadvantaged by a lack of health infrastructure and services (RHIH 2018; Monnat and Rigg 2016). This may mean that rural communities benefit less from strategies that aim to reduce drug supply (e.g. law enforcement) and demand (e.g. behavioural and pharmaceutical interventions). Anecdotal reports in Australia suggest that small rural communities may be deliberately targeted by drug traffickers (Roche and McEntee 2017), who sell drugs cheaply to establish local demand (Commonwealth of Australia, Department of the Prime Minister and Cabinet 2015). A lack of police resources may enable these marketing strategies to succeed.

Where therapeutic services are available, uptake may be impeded by concerns about preserving anonymity. People who need assistance to deal with substance use may be worried about being readily identified in a small community (RHIH 2018).

Harm minimisation strategies may be deficient too. By way of example, America's federally funded Rural Health Information Hub (2018) has suggested that the risks of drug overdoses may be increased in rural settings because of (a) limited experience among first responders and emergency room staff and (b) lack of access to medications to treat drug dependence, such as methadone for opioid dependence.

It is possible that the economic and social harms of substance use are more concentrated in some rural communities. We have previously suggested that the economic and social burdens of problematic substance use by any one individual are likely to be felt more acutely in a small rural community than in an urban setting (Prichard et al. 2018). For example, in a city the harms caused by one person's substance use might be dispersed over a number of locations. They may lose their job in one suburb (possibly affecting a small business), have a traffic accident in another (requiring health resources) and engage in family violence in the suburb where they live (affecting other people and requiring police resources). In small rural settings, by contrast, it is more likely that these harms will occur within the boundaries of the same community. We will return to this important point again when we discuss WWA research designs in rural settings.

Despite the difficulties faced in some rural areas, comparisons of urban and rural areas present a complex picture – one that is not uniform. For instance, broad patterns in America have indicated that in some parts of the country (northeast and midwest) the poorest health outcomes are found in large metropolitan populations, but in other parts (south and west) they are found in nonmetropolitan county populations (Meit et al. 2014). Community type, rural or urban, per se does not lead to specific types of health disparities (Smith, Humphreys and Wilson 2008).

In the same vein, research on psychoactive substance use – as opposed to health indicators – has found that community type per se is not a strong predictor of substance use (see Milano et al. 2017). Given that community type is just one of a number of variables related to substance use, it is not surprising that comparisons of rural and urban areas have produced mixed and sometimes conflicting findings (Rigg and Monnat 2015; see also Borders and Booth 2007; Coomber et al. 2011).

4.1.3 *Extent of Rural Substance Use*

4.1.3.1 America

One of the most comprehensive recent analyses of substance use patterns in urban and rural settings is that of Mack, Jones and Ballesteros (2017; cf. Borders 2018). They examined American trends over several years using two different sources. They analysed data on self-reported drug use and drug use disorders from NSDUH data between 2003 and 2014 and used data from the National Vital Statistics System Mortality (NVSS-M) on drug-related deaths between 1999 and 2015.

The NVSS-M data were used to produce a binary comparison of metropolitan and nonmetropolitan areas. In 1999 the drug overdose death rate per 100,000 people was 6.4 in metropolitan areas and 4.0 for nonmetropolitan areas. The trend leading up to 2015 showed two things. First, drug overdose death rates have increased in both areas. Second, in 2015 the rate of increase was higher in nonmetropolitan (17.0 deaths per 100,000 people) than in metropolitan (16.2 deaths per 100,000 people) areas.

Mack, Jones and Ballesteros (2017, 3242) point to calculations by the US Centers for Disease Control and Prevention which indicate that the increase in the death rate has been primarily driven initially by the use of diverted pharmaceutical opioids and later by heroin, illicitly manufactured fentanyl, cocaine and methamphetamines.

What did survey data indicate? In the 2012–14 period, the percentage of NSDUH participants who reported using illicit drugs in the past month (called 'recent use') was highest in large metropolitan areas (10.1%), followed by 9.5% for small metropolitan areas and 6.8% for nonmetropolitan areas (Mack, Jones and Ballesteros 2017). When compared against results from the 2003–2005 period, nonmetropolitan drug use increased by 13.3% but less than in small (15.9%) and large metropolitan (21.7%) areas.

Further detail about self-reported recent use of psychoactive substances is provided in Table 4.1 from the 2016 NSDUH (detailed tables) (SAMHSA 2017b) on three subcategories of nonmetropolitan areas: urbanised, less urbanised and completely rural. The 2016 NSDUH tables are based on interviews with 67,942 participants in America, which represents approximately 0.021% of the country's population (325.7 million).

Consistent with the longitudinal trends reported by Mack, Jones and Ballesteros (2017), the prevalence of recent use of any illicit drug was lowest in nonmetropolitan areas.

However, different patterns emerged for the drugs associated with overdose deaths. The highest prevalence of recent methamphetamine use was in less urbanised areas (0.5%). Cocaine use (including crack cocaine) was reported by 0.6% of participants from urbanised areas – which is comparable to small metropolitan areas (0.7%). Likewise, recent use of opioids (heroin or prescription pain relievers) in urbanised (1.2%) and less urbanised areas (1.3%) matched the finding for large metropolitan areas (1.3%).

Table 4.1 also shows that heavy alcohol use is considerably more common in rural settings. In completely rural areas 1.16% of participants reported heavy use compared with respectively 0.19% and 0.24% of participants in large metropolitan and small metropolitan areas. Similarly, recent tobacco use was highest in nonmetropolitan areas (31.6%).

4.1.3.2 Australia

Australian research has produced similar findings to those in the USA. Pharmaceutical opioid overdose deaths increased between 2001 and 2012 (Roxburgh et al. 2017) as did deaths related to methamphetamine consumption (e.g. relating to accidental overdose, disease and suicide)

Table 4.1 Prevalence (%) of Recent (Past Month) Reported Substance Use in 2016 by County Type (USA), Participants Aged 12 years+

	Any Illicit Drug	Methamphetamine	Cocaine*	Opioid**	Heavy Alcohol Use***	Tobacco
Large metro	11.3	0.2	0.8	1.3	0.19	20.7
Small metro	10.4	0.3	0.7	1.5	0.24	25
Nonmetro	8.4	0.4	0.4	1.2	0.32	31.6
Urbanised	8.9	0.2	0.6	1.2	0.52	28.8
Less urbanised	8.3	0.5	0.2	1.3	0.41	33.6
Completely rural	6.8	0	0.3	0.9	1.16	33

* *Cocaine includes crack.*

** Opioids includes heroin or the misuse of prescription pain relievers.

*** Heavy alcohol use is defined as binge drinking on each of five or more days in the past 30 days. Binge drinking means consuming 5+ drinks (males) or 4+ drinks (females) on one occasion.

Source: Adapted from NSDUH 2016 detailed tables (Substance Abuse and Mental Health Services Administration 2017b).

Table 4.2 Prevalence (%) of Reported Substance Use in 2016 by Area (Australia), Participants Aged 14 years+

	Major Cities	Inner Regional	Outer Regional	Remote/Very Remote
Persons	70.0	18.9	8.3	2.7
Tobacco (daily)	10.6	14.9	17.0	20.7
Alcohol (11+ drinks monthly+)	6.3	7.8	9.1	15.0
Use in previous year				
Any illicit drug	15.6	14.9	14.4	24.8
Cannabis	10.4	10.1	9.3	17.0
Ecstasy	2.5	1.5	*1.2	1.9
*Meth/amphetamine***	1.4	1.2	1.6	*3.5
Cocaine	3.2	1.3	*0.7	*0.7
*Painkillers/analgesics & opioids***	3.3	3.6	4.2	*6.6
*Pharmaceuticals***	4.6	4.7	5.3	*8.0

* *Estimate has a relative standard error of 25% to 50% and should be used with caution.*
** For non-medical purposes.
Source: Adapted from AIHW (2017b).

between 2009 and 2015 (Darke, Kaye and Duflou 2017). Unlike America and other countries, Australia does not have an established market in crack cocaine (ACIC 2019) but like America, methamphetamine consumption has increased markedly in recent years (ACIC 2019).

An overview of the prevalence of psychoactive substances use (including misuse of pharmaceuticals) in Australia is provided in Table 4.2. This presents data from the leading survey of substance use in the general population: the National Drug Strategy Household Survey (NDSHS) (AIHW 2017a). Australia also has other long-standing monitoring systems. But it is important to note that these concentrate on capital cities mainly because these systems are not designed to study substance use in the general population (Uporova et al. 2018, 1; Karlsson and Burns 2018, 1; Patterson et al. 2018, 28).

The NDSHS categorises areas into major cities, inner regional, outer regional and remote/very remote. The procedure it employed differs from the NSDUH in some respects:

- Participants are eligible only if they are aged 14 years or older; the minimum age limit for the NSDUH is 12 years (see further AIHW 2017a, 133–141; Kilmer, Reuter and Giommoni 2015, 235, 247).

- While the NDSHS is conducted every two to three years – and not annually like the NSDUH – it samples a larger proportion of the population. In 2016 the NDSHS surveyed 23,772 people, which constitutes approximately 0.098% of the Australian population (24.13 million).
- Unlike the American survey, the NDSHS did not ask its participants about their use of drugs and misuse of pharmaceuticals *in the past month*. Consequently, the Australian figures shown in Table 4.2 – regarding drug use in the past year – are higher than those listed in Table 4.1 (which show use in the past month).

About nine out of every ten survey participants were from major cities or inner regional areas. This is broadly consistent with the distribution of the Australian population (see ABS 2018). The regular use of tobacco and alcohol increased with remoteness – with the lowest reported levels in the major cities and the highest in remote/very remote areas. People in remote/very remote places reported regular use of tobacco (20.7%) and alcohol (15%) at twice the rates of the people in major cities (10.6%, 6.3%).

Major cities recorded the highest prevalence of cocaine and ecstasy use but use of any drug in the last year was highest in remote/very remote areas. This was reported by one-quarter of participants from those locations, compared with 15% in other places. This was driven primarily by the prevalence of cannabis use, which was approximately 70% higher in remote/very remote places. The latter areas also reported higher levels of methamphetamine use, and the misuse of pharmaceuticals, painkillers/analgesics and opioids. However, because of the difficulties associated with recruiting a sufficient number of people in remote/very remote locations, these figures must be interpreted with caution because they have a relative standard error rate of 25–50% (AIHW 2017b).[2]

Roche and McEntee (2017) recently conducted comparative analyses of NDSHS data collected in 2007, 2010 and 2013 and Australian drug treatment data for 2006–07, 2009–10 and 2012–13. Locations were categorised into city, regional and rural areas. A statistically significant increase in methamphetamine-related treatment episodes was found in all areas – although the increase in rural areas was proportionally smaller than in regional areas and cities. However, the NDSHS data showed different statistically significant trends. Compared with city and regional areas, participants from rural areas reported the highest methamphetamine use over the lifetime and in the last 12 months.

It is important to point out that Roche and McEntee (2017) were not able to compare Australia's six states and two territories. Despite the fact that their analyses encompassed three NDSHS datasets (N=71,817), the sample sizes in rural areas did not provide enough statistical power for this level of analysis.

4.2 *Rural Wastewater Monitoring: The Australian Experience*

As flagged at the outset of this chapter, the value of concentrating our discussion upon America and Australia lies in the fact that both these large countries use traditional research methods to monitor substance use across very large geographical areas. The WWA monitoring system that Australia has recently established would be applicable for the examination of rural drug-related issues not only for the American context, but also for other large nations, such as Canada, India, China, Russia and Brazil.

Australia's monitoring system has only just begun to demonstrate the potential utility of WWA in rural areas. But it has made more progress on rural monitoring than the European system operated by SCORE under the aegis of the EMCDDA. This is mainly because of necessity: Australia's 24 million people are spread over 7.7 million square kilometres, which comprises 75% of the landmass of Western Europe.

With sufficient funding it is feasible that SCORE could increase its representation of rural areas in future years to increase confidence that its current samples reflect national, rather than urban trends. Given that the Australian monitoring system is national and longitudinal, it can be differentiated from other multi-sites studies that have been conducted in other regions, such as France (Nefau et al. 2013).

The Australian Criminal Intelligence Commission commenced funding for the National Wastewater Drug Monitoring Program (NWDMP) in 2016. This program builds on the larger-scale sampling undertaken by SewAus,[3] an Australian Research Council Linkage Project led by researchers at The University of Queensland. The NWDMP is a collaboration between the University of South Australia, Adelaide, and the University of Queensland, Brisbane that draws on the strengths of both pioneering teams of WWA in Australia. Both the NWDMP and SewAus are supported by local water authorities, which collect and transport samples to the university laboratories, following accepted protocols (see Chapter 2).

4.2.1 *Findings of the Australian National Wastewater Drug Monitoring Program in 2017*

The NWDMP reports are published three times each year on data from approximately 50 sampling sites (the number fluctuates slightly) that collectively service over 12 million people, more than half the Australian population (ACIC 2019). The number of samples collected from each site in each data collection wave ranges from four to seven. The analysis of the samples and writing of the reports takes three to four months. This demonstrates the speed with which WWA data can be collated, analysed and reported – astonishing for research conducted at a national level.

In terms of national trends between 2016 and 2017, the NWDMP estimated that 8.4 tonnes of methamphetamine was consumed annually in Australia. This is in addition to 3.1 tonnes of cocaine, 1.3 tonnes of ecstasy and 0.8 tonnes of heroin (ACIC 2018). Comparing this against national seizure data, the ACIC (2018) calculated that law enforcement agencies seized approximately:

- 25% of the heroin estimated to meet national demand; and
- 40% of the methamphetamine needed to meet national demand.

The amount of ecstasy seized equalled annual consumption. And seizures of cocaine exceeded the total weight required for national demand.

What did the NWDMP analyses reveal about rural substance use? The NWDMP uses a binary variable for remoteness: capital cities and all other areas, which are defined as 'regional areas' (ACIC 2018, 1). Because this categorisation is so broad – for example, encompassing large suburbs – it is minimally useful in examining rural trends. We will return to this limitation presently.

The NWDMP reports identify the trends of the capital cities. The locations of the other sites are de-identified, a practice that is consistent with WWA ethical guidelines (Prichard et al. 2016). However, the results from each location are revealed in confidence to the participating municipalities. This means that local authorities have access to trends in their own community.

Figure 4.1 presents findings from the NWDMP on methamphetamine and fentanyl for the Australian States and Territories.[4] We chose these substances as examples because of the harms associated with their misuse.

Given the limitations of the wide definition of 'regional', no conclusions can be drawn from these data with respect to rural substance use in Australia. Compared with regional areas, Australian capital cities have a lower average consumption rate of fentanyl. But, within 'regional areas', it is unclear whether fentanyl consumption differs in any way between large suburbs and very remote places.

Nonetheless, the data show the potential utility of WWA for rural areas. First, they show how readily drug consumption trends can be compared across multiple sites within one country. This feature of WWA can assist to identify specific 'hot-spots' or locations where substance use is unusually high.

Two good examples can be seen in Figure 4.1. The first is the estimated consumption of methamphetamine in site 066, in Victoria (VIC). This location recorded methamphetamine consumption that was two to three times

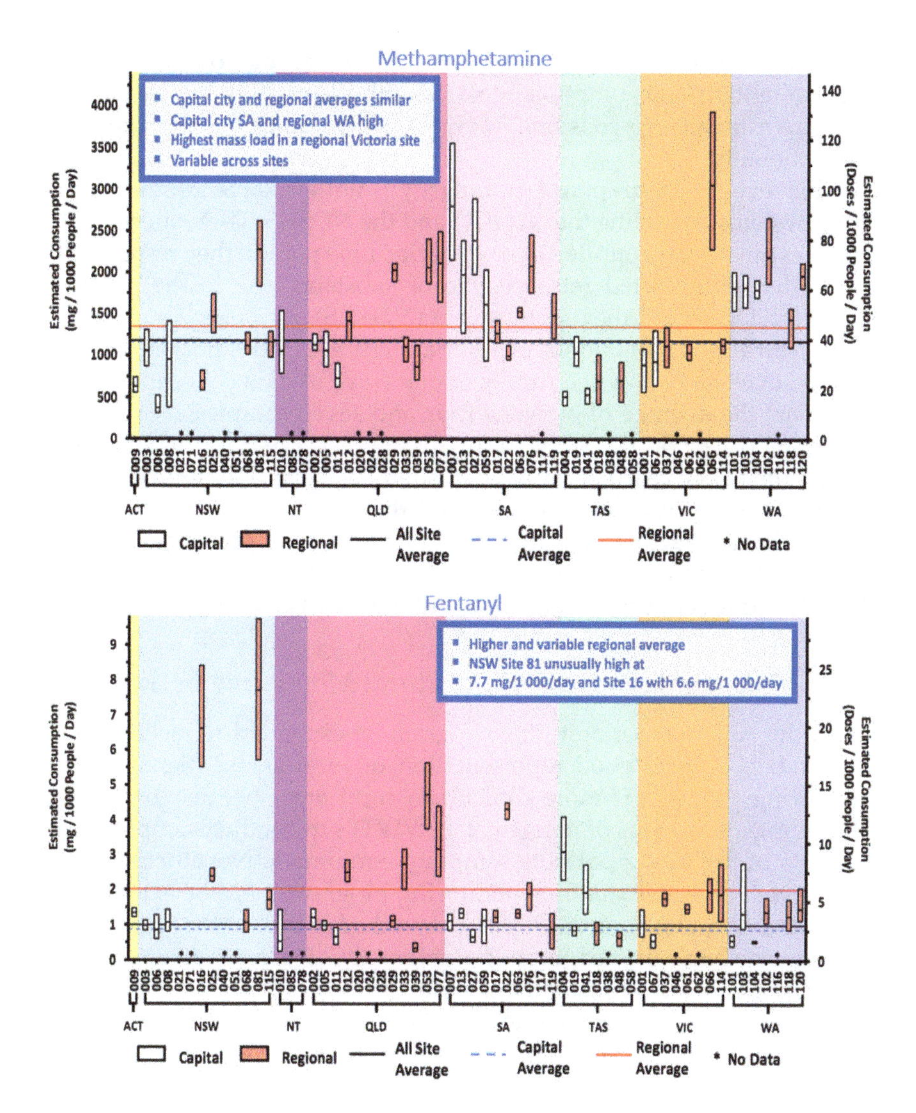

Figure 4.1 Estimated Consumption for December 2017 in Mass Consumed per Day (Left Axis) and Doses per Day (Right Axis) per Thousand People, by Jurisdiction

Capital cities are presented in white and all other areas ('Regional') are presented in red.

Source: ACIC (2018, 25, 29).

higher than all other Victorian sites. The second example concerns the estimates of fentanyl consumption in New South Wales (NSW). Sites 016 and 082 were estimated to consume in excess of five times the average of all sites nationally.

This level of geographical granularity is not available in other monitoring systems, including the NSDUH and the NDSHS. These surveys also cannot estimate consumption over a specific time period; they respectively record the self-reported retrospective of substance use in the previous month (NSDUH) and the previous year (NDSHS).

A critical point to underscore is that Figure 4.1 does not present any patterns over time, only the results of one wave of data collection (which represents the average of between four and seven samples at each site). Since data collection takes places three times each year for regional sites and six times per year for capital city sites, each site will in time be able to develop relatively fine-grained trend data. If the program continues, within 10 years each regional site could have 30 data points and 60 data points for the capital city sites. By contrast, over the same time period the NSDUH would have ten data points and the NDSHS three or four.

4.2.2 *Potential Improvements for Rural Monitoring in Australia*

Could the Australian monitoring program be expanded to include more sites and to achieve good representation of rural areas? Managing the monitoring program is more difficult in rural areas because of, among other things, shortages of personnel at WWTPs to conduct sampling and the costs of providing portable sampling equipment. Nonetheless, there are reasons to believe that the monitoring program could be extended to more rural areas. We base this argument on the fact that, at the time of writing, SewAus had sampled from approximately 130 sites across Australia, including the 50 sites already included in the NWDMP.

Tables 4.3 and 4.4 provide information about the numbers of sites in each jurisdiction, the approximate population size of each catchment, and how the sites could be classified according to the Australian remoteness code (major city, inner regional, outer regional, remote and very remote).

Catchment population is estimated by Australian Bureau of Statistics Census 2016. Population estimates are missing for an additional 20 sites. Actual population estimates are not shown to protect the anonymity of the communities where sampling occurs.

Table 4.3 shows that the NWDMP has good, albeit uneven, representation in all jurisdictions in proportion to their total populations. Eighteen of the sites service large catchments of over 200,000 people, the largest serving approximately 2.2 million people. Half of the sites (65) service populations of between 5,000 and 80,000. Sixteen of the smallest sites treat wastewater from less than 5,000 people.

Table 4.3 SewAus Wastewater Sampling Sites 2018, by Jurisdiction and Population (*N*=110)

Jurisdiction (Total Population, Thousands)	Sewerage Catchment Population Bracket* (Thousands)						
	<5	>5–20	>20–80	>80–200	>200–500	>500–2,200	N
Aus. Capital Territory (397)	–	–	–	–	1	–	1
New South Wales (7,480)	–	3	6	4	2	3	18
Northern Territory (229)	2	1	4	–	–	–	7
Queensland (4,703)	11	12	14	5	3	1	46
South Australia (1,677)	1	3	3	1	1	1	10
Tasmania (509)	2	3	4	–	–	–	9
Victoria (5,926)	–	3	4	1	1	2	11
Western Australia (2,474)	–	2	3	–	1	2	8
Total	**16**	**27**	**38**	**11**	**9**	**9**	**110**

Source: Adapted from SewAus.

Table 4.4 SewAus Wastewater Sampling Sites 2018, by Remoteness and Average Population

	Sites	Missing Population Data	Average Catchment Population (Excluding 20 Missing Cases)
Major cities	41	2	387,303
Inner regional	32	5	27,622
Outer regional	40	5	19,387
Remote	11	2	8,855
Very remote	6	6	–
Total	**130**	**20**	–

Source: Adapted from SewAus.

Table 4.4 aggregates the sites according to their remoteness code, which is the same coding used by the Australian general population survey, the NDSHS.

The 41 sites located in major cities clearly service many times the number of people served by sites in other settings, with an average catchment population of 387,000. Nonetheless, wastewater sampling is occurring in 89 locations in regional and remote parts of Australia.

At the time of writing, population estimates were not available for the six very remote sites and two remote ones. However, the remaining nine remote community's average population was 8,855 people. Reference to raw data shows that three of the remote communities had populations under 2,000 people. A further three have populations between 5,000 and 9,000. The last three service between 10,000 and 20,000 – two sites in Queensland and one in the Northern Territory.

As discussed in Chapter 2, WWA calculations are based on assumptions relating to the 'standard' metabolic functioning of the human body. Although the majority of individuals metabolise drugs in a 'standard' fashion there are some individuals who metabolise drugs to a greater extent than 'standard' and others who metabolise drugs to a lesser extent than 'standard'. In large populations, these individuals do not have a significant statistical impact upon the reliability of drug consumption estimates. In small populations, by contrast, the statistical influence of individuals may lead to greater uncertainty in drug consumption estimates. Therefore, for values estimated from small populations larger ± values are likely applicable. To date the WWA field has not developed clear guidelines about the empirical validity of analysing wastewater data from small communities. Castiglioni and Lor (2016, 36) suggested that 10,000 might represent the lower level for reliability.

Applying this guide to the SewAus sites, only three of the remote locations would be higher than the 10,000-population reliability guideline. All of the sites in major cities, inner regional and outer regional places exceed this number.

This means that each of these eligible sites could be compared against each other with an acceptable level of reliability three times per year. This would be extremely difficult to achieve using survey methods in any setting. But the challenges in smaller rural communities would be particularly acute; the problems associated with recruiting participants in such settings would be compounded by survey fatigue. In other words, even if the barriers to recruitment, discussed above in 4.1.1, could be overcome for the first data collection survey, it is unlikely that a sufficient sample size could be recruited every four months or so.

4.2.3 *Limitations Relating to Jurisdictional Comparisons of Remoteness*

Could Australia's six states and two territories be compared according to the five remoteness categories (major cities, inner regional, outer regional, remote and very remote)? Earlier (4.1.3) we noted that, due to an insufficient sample size in some rural areas, such an analysis could not be achieved even when data from three of Australia's general population surveys was combined (Roche and McEntee 2017). Table 4.5 presents current SewAus sites in each jurisdiction ranked by remoteness categories.

There are good prospects for comparing jurisdictions according to major cities, inner regional and outer regional areas. However, clearly

Table 4.5 SewAus Wastewater Sampling Sites 2018, by Remoteness and Jurisdiction

	Major City	Inner Regional	Outer Regional	Remote	Very Remote	*N*
ACT*	2	–	–	–	–	2
NSW	14	6	3	–	–	23
NT	–	–	4	3	–	7
QLD	14	10	19	4	–	47
SA	4	2	3	1	–	10
TAS	–	6	3	–	–	9
VIC	4	7	5	–	–	16
WA	3	1	3	3	6	16
Total	**41**	**32**	**40**	**11**	**6**	**130**

* *The Australian Capital Territory is solely the location of the national capital, Canberra, and consequently the whole jurisdiction falls into the category 'major city'.*

Source: Adapted from SewAus.

WWA data are not collected from many very remote locations. Six are in Western Australia, but as indicated in Table 4.4, the population sizes of these catchments are presently not known so it is unclear whether sampling reliability guideline requirements could be met. Better prospects exist for the comparison of 'remote' locations by jurisdiction. We highlighted above that 3 of the 11 remote sites have sufficient population sizes for WWA.

In short, with publicly available information we were not able to determine whether *any* very remote communities exist in Australia where WWA could be reliably conducted. It is also unclear whether enough suitable 'remote' sites exist with which to make meaningful comparisons of the jurisdictions.

This area of uncertainty highlights an important potential limitation of WWA. It means that data on substance use in some remote and very remote locations can *only* be captured with traditional methods. The other implication is that WWA may not be able to circumvent the sorts of problems encountered by Roche and McEntee (2017), which means that *jurisdictional comparisons* of remote and very remote locations may not be achievable with WWA.

4.3 Other Applications of Wastewater Analysis for Rural Settings

How else might WWA be of use in studying drug use in rural communities? In this section we explore the possible value of short-term WWA studies to gauge the prevalence of consumption. We also consider the possibility for conducting quasi-experimental research in rural communities, assuming the availability of suitable sampling sites and WWTP catchments that service a population of 10,000 or more people.

4.3.1 Short-Term Wastewater Studies in Rural Communities

Markets for illicit drugs and diverted pharmaceuticals are complex and often unpredictable because many factors can influence supply and demand. As a case in point, the growth in American domestic demand for fentanyl, heroin, methamphetamine and other substances (see 4.1.3) was not foreseen 20 years ago.

Because of the importance of being able to detect changes in market behaviour, many countries have some form of early warning system (e.g. EMCDDA 2012). In America the National Drug Early Warning System draws on a wide range of experts and data sources to detect emerging drug problems (NIDA 2018). Early warning capabilities in Australia are provided through three main systems: the Illicit Drug Reporting System

(IDRS) (Uporova et al. 2018); the Ecstasy and Related Drug Reporting System (EDRS) (Karlsson and Burns 2018); and Drug Use Monitoring Australia (DUMA) (Patterson et al. 2018). The IDRS and EDRS use a combination of surveys of people who recently used illicit drugs, key informants from health and law enforcement professionals, data from forensic laboratories and so forth. Like a now defunct American program (see Kilmer, Reuter and Giommoni 2015), DUMA analyses survey data and urine samples collected from arrestees. As noted above in 4.1.3, these three systems operate only in capital cities.

Elsewhere we have suggested that stand-alone and time-limited WWA studies may be useful for rural communities that lie in a 'data shadow' – in other words, in an area where there are few metrics on substance use because the communities lie outside of the boundaries of early warning systems and general population surveys (Prichard et al. 2018, 195).

Broadly speaking, it appears that health and law enforcement professionals are well-equipped to detect the emergence of substance misuse in most small communities once the harms associated with the misuse appear, including, for example, adverse health effects and violence related to alcohol or illicit drugs (see Chapter 1). However, situations can occur where expert opinion is *inconsistent*. Community and public anxiety can be heightened when the specific substance has high levels of social notoriety. One such substance is crystalline methamphetamine (ice) which is associated with health problems (McKetin et al. 2010; Darke et al. 2008; Kaye et al. 2007; Degenhardt, Roxburgh and McKetin 2007) and hostility (McKetin et al. 2016).

The case study we explored in Australia concerned ice use in a small coastal town in rural Tasmania, which is 85 km from the nearest population centre. The town attracted national media attention because health professionals in the town estimated that 10% of residents were "addicted" to ice (Prichard et al. 2018, 197). Other stakeholders suggested that such claims were unfounded. A lack of data from traditional monitoring systems precluded an accurate report on the extent of ice use in the town.

WWA may play a valuable role in situations such as this where qualitative data and event data (see 1.3) are conflicting – *providing that reliable samples can be collected.* Local water authorities could, following instruction, collect samples from the community WWTP over several consecutive days. Appropriately treated samples could be flown to a domestic laboratory and results potentially provided within a fortnight. The WWA data could not be used to determine how many people in a community used a substance (see 2.5) but it could compare consumption levels in the community with those in other regional and national areas. Of course, comparisons that included any very small communities (e.g. those with populations below the 10,000 members threshold proposed by Castiglioni and Lor (2016)) would require caution because of the problems we discussed earlier.

To picture how this might look, it may be useful to refer again to Figure 4.1. Whatever the outcome, the findings would assist the situation of drug use in a rural community within a broader context. Very low estimates, such as those found for fentanyl consumption in site 039, would not support concerns about its consumption. Conversely, such fears could indeed be supported with very high estimates, such as those found for fentanyl consumption in sites 016 and 082. This result would support the case for immediate attention.

A final point is that, if anxieties about a rural community warranted special attention by researchers, WWA would have practical advantages over survey methods. The national media coverage that substance use in the small community in Tasmania received would present serious problems for conducting a survey of drug use. Recruitment would be more difficult because of heightened concerns over anonymity. And there could be greater incentives for participants to deliberately under-report substance use because of, for instance, fears of increased police activity that could affect them personally.

Event data, such as drug-related arrests and overdoses, would definitely be important to monitor. But as a means of research, in this instance, the numbers of cases may be too small to detect trends and these events are reported with lengthy delays. In other words, relying on event data to gauge whether a problem exists requires the community to wait for drug-related harms to occur. In contrast to surveys and event data, WWA could be conducted efficiently without exacerbating community tensions. In theory WWA could detect an emerging substance use problem before it was evident in rates of substance use-related harms (see 1.2).

4.3.2 *Quasi-Experimental Studies*

Randomised controlled trials (RCT) provide the most robust empirical design for causal inference. There are examples of their use to examine the effectiveness of substance use strategies within health (see Babor et al. 2010a) and law enforcement (e.g. Sherman et al. 2015). However, they are costly, can generate ethical complexities (Babor et al. 2010a), and the way that governments usually implement drug strategies tends to preclude the use of RCTs (Hall 2018).

Opportunities for quasi-experimental designs arise when conditions change, for example because of the implementation of a new strategy, or because of unplanned events or factors. Drug use in the area or group subject to the change can be compared with that in similar groups or areas where no such change has occurred. Such designs can provide useful information for policy evaluation and causal inference, although quasi-experiments cannot control extraneous variables (e.g. selection bias) as effectively as RCTs. Hall (2018) has argued that in the context of substance use, analyses of the effects of changes in conditions tend to be retrospective and sometimes limited because they draw on data that were collected for different purposes, such as 'event data' gathered by health and law enforcement agencies (see 1.3.1).

Discussion on the topic of research designs for substance use has paid very little attention to using the new opportunities that WWA might provide. The idea of attempting, for example, a quasi-experimental study with WWA in urban and suburban settings is problematic, but not impossible. Much would depend on the nature of the change and its likely effects.

For example, the Australian Government has implemented a stringent tobacco demand reduction strategy. This involves, among other things, a series of taxation increases over a period of years. Because of the breadth and duration of the Government's intervention, WWA samples from metropolitan WWTPs could be used to evaluate aspects of the strategy. The effectiveness of this strategy can be assessed by data from the NDSHS. This general population survey provides valuable person-centric data and is conducted every two to three years (see 4.1.3). However, as recently demonstrated by Mackie et al. (2019), WWA can provide useful additional time-sensitive evidence – such as the effects of taxation increase. A metropolitan WWA study may also be useful to evaluate supply reduction strategies that targeted, for instance, cargo routes to island cities, such as Honolulu and Hobart, Tasmania.

WWA might be of limited utility when changes – be they caused by government strategies or unplanned factors – are localised to a particular city or suburb. This is because of the boundaries of the lack of alignment between small areas and catchments of WWTPs and population movement into catchment areas. For example, if the suburb was one of several suburbs that all fell within one catchment for a large WWTP, then multiple population groups would be captured and WWA metrics could not isolate drug use in the suburb in question.

Conversely, if the suburb of interest happens to be serviced by a dedicated WWTP, population movement between the catchments could not be controlled for in WWA research. Individuals could consume drugs in a different area but move *into* the suburb and then excrete the drug biomarkers in their urine – producing an increase in drug consumption in the suburb of interest. Equally, individuals could consume drugs within the suburb but move *out* into a different area and excrete the biomarkers in another catchment – producing a falsely low result.

Similar issues may affect WWA studies that examine the relationship between the consumption of substances and health harms. As we discussed earlier (4.1.2), in urban or suburban settings harms may be dispersed geographically and occur in places other than where the substance is used. Of course, dispersion could affect *any* study that examined harms with event data, such as drug-related traffic accidents, overdoses and arrests. But our point is that dispersion presents a further complexity for WWA in metropolitan areas.

Rural towns provide opportunities for WWA researchers to mitigate these problems. In rural settings there is a greater likelihood of a 'continuity of place' between the locations of (a) substance consumption, (b) substance excretion and (c) substance-related harms (Prichard et al. 2018, 204).

From an empirical perspective, the best sites for a rural WWA study would have a single sewerage catchment, few dwellings connected to septic systems, and a low level of population movement. Assessments of the degree of population movement would need to account for multiple circumstances, e.g. drive-time to the nearest major population centre, numbers of tourists, whether the town lies on a major highway and so forth. Seasonal fluctuations will also be relevant. For instance, population movement may be lowest in winter.

Could a retrospective WWA study be conducted in a rural site with optimal conditions – in other words evaluating the effect of a 'change' after the event? This would entirely depend on whether samples had already been collected from the township, which might occur as part of national monitoring.

We think that far greater potential lies in planned intervention studies that evaluate the effects of implementing government strategies (Prichard et al. 2018). Our experience in Australia suggests that there are good opportunities for WWA researchers to collaborate with local rural authorities and government agencies from health and law enforcement portfolios. WWA researchers' two main roles would be to provide expert advice on the suitability of rural towns for WWA and to analyse substance use in wastewater at the site.

The most basic research design would be single case A-B study, where A represented a period of baseline data gathering on substance use behaviours in the rural town and B represented the period of the intervention. The length of WWA data collection in both period A and period B would depend on several factors. The quality of the sampling environment is one important consideration. Where sampling involves few uncertainties (e.g. because of accurate population data, the quality of the sewers, etc.), fewer samples will be required for robust assessment of differences between A and B.

The timing of A and B will partly depend on the nature of the intervention under examination. Some interventions, like supply reduction strategies, might be expected to have effects relatively quickly while others, such as educational demand reduction strategies, may be expected to influence behaviours over a longer time frame.

An A-B-A-B design may be feasible, but this again depends on the nature of the intervention. The repetition of the baseline (A) and intervention (B) periods would provide an internal control of sorts. If WWA data indicated that changes occurred in the consumption of substances after *both* B periods, researchers would have greater confidence that there was a causal link between the intervention and the behavioural change.

More sophisticated single case experimental designs are used in other fields. Gast, Ledford and Severini (2018) describe means to assess the effects of two interventions independently (e.g. A-B-C-B-C) and in combination (e.g. A-B-BC-B-BC). Whether such designs could be implemented in a rural WWA study is unclear. However, the WWA field can only benefit from increasing use of the variety of approaches that researchers utilise in other contexts.

With more planning, rural research could incorporate other towns to act as control groups, monitoring them over the same period as the experimental site (Prichard et al. 2018). Attempts could be made to match the experimental and control sites on geographic, social and economic profiles. Control sites would assist researchers to assess whether any changes observed in the experimental site were due to external factors, including fluctuations in regional markets for illicit drugs or the effects of economic and climatic factors such as unemployment and drought.

4.3.3 Ethical Considerations

As highlighted in Chapter 2 (2.3), WWA research raises special ethical considerations in some circumstances. Although researchers hold an obligation to ensure that they are abiding by the ethical procedures stipulated by their institution, generally ethical risks can be managed by de-identifying locations and adopting protocols for interacting with media. In human research ethics the threshold of 'harm' is low; it can encompass distress and embarrassment and certainly covers stigmatisation and economic harm.

Rural communities may be more vulnerable to 'distress' because they are made up of a smaller group of people who will feel more easily identifiable than residents in large cities. Risks of economic harm from the misreporting of WWA studies are also arguably greater in rural contexts. This may be particularly important for rural communities that are hoping to attract new residents and for towns that rely on tourism.

In regards to stigmatisation, in our view ethical risks are particularly acute for rural settings that include Indigenous communities. We would recommend that WWA researchers seek the oversight of human research ethics committees (HREC) for studies that focussed on particular towns, as discussed in 4.3.1 and 4.3.2. This is because of the likely need for:

- An appreciation of the historic stigmatisation of Indigenous people;
- Indigenous stakeholders to approve and collaborate in the research design; and
- Special consideration of the strategies proposed by health or law enforcement agencies to be included in quasi-experimental designs. By way of example, an HREC might not approve a WWA intervention study that examined the effect of increasing arrests for drug possession.

4.4 *Conclusion*

Patterns of substance use within national borders are often complex. Examining these patterns across high-density urban settings and remote and sparsely populated communities is very challenging, particularly in geographically large nations. Scholars in epidemiology and criminology have ignored the need for research on substance use outside of urban and suburban settings. In part this seems to have been motivated by idealised notions of 'rural' life, notwithstanding the fact that the distinction between rural and non-rural is hard to define within cultures. But it is also the case that studying substance use in rural places is difficult, slow and expensive. Problems in recruiting participants for surveys can be complicated in rural areas because residents are circumspect about 'outsiders' and uncertain about protecting their anonymity. For these reasons ongoing systems that monitor substance use tend to concentrate on metropolitan areas.

Australia's WWA monitoring system is quite new. The federally funded system reports on 50 sites three times per year and 20 of these sites six times a year. These sites collectively reflect substance use behaviours of over 50% of the Australian population. This system is subject to the limitations of WWA that we discussed in Chapter 2 (2.5). However, its scale and temporal sensitivity provide advantages in monitoring substance use in the general population. By comparison, the Australian general population survey (NDSHS) collects data on less than 1% of the population every two to three years. The American survey (NSDUH) is conducted annually and accesses approximately 0.021% of the country's population.

SewAus has collected WWA samples from a total of 130 sites across Australia, including the 50 sites in the monitoring system. Data from SewAus provides a useful indication of the capability of WWA to monitor substance use in rural places. The most striking feature of results to date is that WWA can compare sites against each other individually. Among other things, it means that WWA may help to identify particular hotspots for consumption of particular substances. If monitoring continues, the compilation of WWA findings will form a useful dataset for longitudinal analyses.

Situations can arise in rural communities where a time-limited, single WWA may help to resolve disagreements about the levels of consumption of psychoactive substances in the community. Such situations may not arise frequently, but for rural communities in data shadows where there are no survey data, WWA may be helpful.

Perhaps the most empirically important aspect of our discussion concerned quasi-experimental studies in rural areas. Paradoxically, the WWA method means that in some circumstances remoteness – which normally disadvantages rural communities in terms of research – may be an empirical advantage. In the right conditions, remoteness may provide researchers with confidence that there is a continuity of place; that is, that substance use, the excretion of the substance and associated harms mainly occur in one catchment. Such conditions are less likely to be met in a non-rural setting.

Providing ethical considerations for the field are observed, WWA researchers could potentially collaborate with government agencies to assess the efficacy of strategies to reduce drug use using quasi-experimental methods in rural townships – either with single case designs or by including control groups. This type of design has not been possible to date because the traditional sources of data on substance consumption lack the accuracy and the temporal granularity of WWA. We intend to pursue the application of WWA in this way. If successful, this approach has the potential to provide high-quality evidence about the efficacy of strategies that target rural communities.

Notes

1. Afghanistan, Austria, Chile, Colombia, Croatia, Finland, Hungary, Iceland, India, Kenya, Mauritius, Russia, Spain, Sri Lanka, Sweden, Switzerland, Tanzania, Turkey, Ukraine and the USA (Milano et al. 2017, 1803).
2. See Table 8.1: Drug use by ASGS remoteness areas, people aged 14 years or older, 2010 to 2016 (per cent) in AIHW (2017b).
3. SewAus is funded by the Australian Government through the Australian Research Council.
4. Australian Capital Territory (ACT), New South Wales (NSW), Northern Territory (NT), Queensland (QLD), South Australia (SA), Tasmania (TAS), Victoria (VIC) and Western Australia (WA).

References

Australian Bureau of Statistics. 2018. Correspondence, 2017 Locality to 2016 Remoteness Area. www.abs.gov.au/AUSSTATS/abs@.nsf/DetailsPage/1270.0.55.005July%202016?OpenDocument (accessed 15 June 2018).

Australian Criminal Intelligence Commission. 2018. National Wastewater Drug Monitoring Program – *Report 5*. www.acic.gov.au/sites/default/files/nwdmp5.pdf?v=1564718845.

Australian Criminal Intelligence Commission. 2019. *National Wastewater Drug Monitoring Program – Report 7*. www.acic.gov.au/sites/default/files/2019/06/nwdmp7_140619.pdf?v=1560498324

Australian Institute of Health and Welfare. 2017a. *National drug strategy household survey 2016: Detailed findings.* Drug Statistics series no. 31. Cat. no. PHE 214. Canberra: Australian Institute of Health and Welfare.

Australian Institute of Health and Welfare. 2017b. *National drug strategy household survey 2016: Detailed findings.* (*Data tables: Chapter 8 Specific population groups*). www.aihw.gov.au/reports/illicit-use-of-drugs/2016-ndshs-detailed/data#page2 (accessed 6 June 2018).

Babor, T. F., J. P. Caulkins, G. Edwards et al. 2010a. *Drug policy and the public good.* Oxford: Oxford Univ. Press.

Barclay, E., and J. F. Donnermeyer. 2002. Property crime and crime prevention on farms in Australia. *Crime Prevention and Community Safety* 4:47–61.

Barclay, E., J. F. Donnermeyer, and P. C. Jobes. 2004. The dark side of *Gemeinschaft*: Criminality within rural communities. *Crime Prevention and Community Safety* 6:7–22.

Barclay, L., A. Phillips, and D. Lyle. 2018. Rural and remote health research: Does the investment match the need?. *The Australian Journal of Rural Health* 26:74–9.

Borders, T. F. 2018. Portraying a more complete picture of illicit drug use epidemiology and policy for rural America: A competing viewpoint to the CDC's *MMWR* Report. *The Journal of Rural Health* 34:3–5.

Borders, T. F., and B. M. Booth. 2007. Rural, suburban, and urban variations in alcohol consumption in the United States: Findings from the National Epidemiologic Survey on Alcohol and Related Conditions. *The Journal of Rural Health* 23:314–21.

Bureau of Infrastructure, Transport and Regional Economics (BITRE). 2017. *Yearbook 2017: Australian Infrastructure Statistics, Statistical Report*. Canberra: Bureau of Infrastructure, Transport and Regional Economics

Carrington, K., J. F. Donnermeyer, and W. S. DeKeseredy. 2014. Intersectionality, rural criminology, and re-imaging the boundaries of critical criminology. *Critical Criminology* 22:463–77.

Castiglioni, S., and E. G. Lor. 2016. Target drug residues in wastewater. In *Assessing illicit drugs in wastewater: Advances in wastewater-based drug epidemiology*, ed. S. Castiglioni, 35–43. Luxembourg: Publications Office of the European Union.

Castiglioni, S., and L. Vandam. 2016. A global overview of wastewater-based epidemiology. In *Assessing illicit drugs in wastewater: Advances in wastewater-based drug epidemiology*, ed. S. Castiglioni, 45–54. Luxembourg: Publications Office of the European Union.

Center for Behavioral Health Statistics and Quality. 2017. *2016 National survey on drug use and health: Detailed tables*. Rockville, MD: Substance Abuse and Mental Health Services Administration.

Central Pollution Control Board. 2015. Inventorization of sewage treatment plants. https://nrcd.nic.in/writereaddata/FileUpload/NewItem_210_Inventorization_of_Sewage-Treatment_Plant.pdf.

Cerdá, M., M. Wall, K. M. Keyes, S. Galea, and D. Hasin. 2012. Medical marijuana laws in 50 states: Investigating the relationship between state legalization of medical marijuana and marijuana use, abuse and dependence. *Drug and Alcohol Dependence* 120:22–7.

Commonwealth of Australia, Department of the Prime Minister and Cabinet. 2015. *Final report of the National Ice Taskforce*. Canberra: Commonwealth of Australia, Department of the Prime Minister and Cabinet.

Coomber, K., J. W. Toumbourou, P. Miller, P. K. Staiger, S. A. Hemphill, and R. F. Catalano. 2011. Rural adolescent alcohol, tobacco, and illicit drug use: A comparison of students in Victoria, Australia, and Washington State, United States. *The Journal of Rural Health* 27:409–15.

Darke, S., S. Kaye, and J. Duflou. 2017. Rates, characteristics and circumstances of methamphetamine-related death in Australia: A national 7-year study. *Addiction* 112:2191–201.

Darke S., S. Kaye, R. McKetin, and J. Duflou. 2008. Major physical and psychological harms of methamphetamine use. *Drug and Alcohol Review* 27:253–62.

Degenhardt L., A. Roxburgh, and R. McKetin. 2007. Hospital separations for cannabis- and methamphetamine-related psychotic episodes in Australia. *The Medical Journal of Australia* 186:342–45.

Delahunty, B., and J. Putt. 2006. *The policing implications of cannabis, amphetamine and other illicit drug use in Aboriginal and Torres Strait Islander communities.* Canberra: National Drug Law Enforcement Research Fund.

Discover Water, United Kingdom. 2018. Industry dashboard. www.discoverwater.co.uk/ (accessed 23 May 2018).

European Monitoring Centre for Drugs and Drug Addiction. 2012. *Early warning system: National profiles*. Luxembourg: Publications Office of the European Union.

European Monitoring Centre for Drugs and Drug Addiction. 2019. *Perspectives on drugs: Wastewater analysis and drugs: A European multi-city study*. www.emcdda.europa.eu/system/files/publications/2757/POD_Wastewater%20analysis_update2019.pdf.

Gast, D. L., J. R. Ledford, and K. E. Severini. 2018. Withdrawal and reversal designs. In *Single case research methodology: Applications in special education and behavioral sciences*, ed. J. R. Ledford, and D. L. Gast, 215–38. New York: Routledge.

Gfroerer, J. C., S. L. Larson, and James D. Colliver. 2007. Drug use patterns and trends in rural communities. *The Journal of Rural Health* 23:10–5.

Hall, W. 2018. The future of the international drug control system and national drug prohibitions. *Addiction* 113:1210–23.

Hedegaard, H., M. Warner, and A. M. Miniño. 2017. Drug overdose deaths in the United States, 1999–2015. *NCHS Data Brief* (273):1–8

Hogg, R. 2011. Governing crime at a distance: Spatiality, law and justice. *Current Issues in Criminal Justice* 22: 361–77.

Jan, A., and N. Khan. 2015. Problems associated with samples collection from people with different ages in rural communities of Pakistan. *The South Asian Journal of Medicine* 1:17–20.

Karlsson, A., and L. Burns. 2018. *Australian Drug Trends 2017. Findings from the Illicit Drug Reporting System (IDRS)*. Australian Drug Trend Series. No. 181. Sydney: National Drug and Alcohol Research Centre, University of New South Wales.

Kaye, S., R. McKetin, J. Duflou, and S. Darke. 2007. Methamphetamine and cardiovascular pathology: A review of the evidence. *Addiction* 102:1204–11.

Kilmer, B., P. Reuter, and L. Giommoni. 2015. What can be learned from cross-national comparisons of data on illegal drugs?. *Crime and Justice* 44:227–96.

Mack, K. A., C. M. Jones, and M. F. Ballesteros. 2017. Illicit drug use, illicit drug use disorders, and drug use overdose deaths in metropolitan and nonmetropolitan areas – United States. *American Journal of Transplantation* 17:3241–52.

Mackie, R. S., B. J. Tscharke, J. W. O'Brien, P. M. Choi, C. E. Gartner, K. V. Thomas, and J. F. Mueller. 2019. Trends in nicotine consumption between 2010 and 2017 in an Australian city using the wastewater-based epidemiology approach. *Environment International* 125:184–190.

McKetin, R., S. Dawe, R. A. Burns et al. 2016. The profile of psychiatric symptoms exacerbated by methamphetamine use. *Drug and Alcohol Dependence* 161:104–9.

McKetin, R., K. Hickey, K. Devlin, and K. Lawrence. 2010. The risk of psychotic symptoms associated with recreational methamphetamine use. *Drug and Alcohol Review* 29:358–63.

Meit, M., A. Knudson, T. Gilbert et al. 2014. The 2014 update of the rural-urban chartbook. https://ruralhealth.und.edu/projects/health-reform-policy-research-center/pdf/2014-rural-urban-chartbook-update.pdf.

Milano, G., E. Saenz, N. Clark et al. 2017. Report on the international workshop on drug prevention and treatment in rural settings organized by United Nations Office on Drugs and Crime (UNODC) and World Health Organization (WHO). *Substance Use & Misuse* 52:1801–7.

Miller, P. G., K. Coomber, P. Staiger, L. Zinkiewicz, and J. W. Toumbourou. 2010. A review of rural and regional alcohol research in Australia. *The Australian Journal of Rural Health* 18:110–7.

Monnat, S. M., and K. K. Rigg. 2016. Examining rural/urban differences in prescription opioid misuse among US adolescents. *The Journal of Rural Health* 32:204–18.

National Institute on Drug Abuse (NIDA). 2018. National Drug Early Warning System. www.drugabuse.gov/related-topics/trends-statistics/national-drug-early-warning-system-ndews (accessed 22 May 2018).

Nefau, T., S. Karolak, L. Castillo, V. Boireau, and Y. Levi. 2013. Presence of illicit drugs and metabolites in influents and effluents of 25 sewage water treatment plants and map of drug consumption in France. *Science of the Total Environment* 461:712–22.

Patterson, E., T. Sullivan, A. Ticehurst, and S. Bricknell. 2018. *Drug use monitoring in Australia: 2015 and 2016 report on drug use among police detainees*. Statistical Report 04. Australian Institute of Criminology. https://aic.gov.au/publications/sr/sr4.

Pierce, C. S., and E. Scherra. 2004. The challenges of data collection in rural dwelling samples. *Online Journal of Rural Nursing and Health Care* 4, no. 2 (December): 25–30. https://rnojournal.binghamton.edu/index.php/RNO/article/view/197.

Prichard, J., W. Hall, E. Zuccato et al. 2016. *Ethical research guidelines for sewage epidemiology*. www.emcdda.europa.eu/system/files/attachments/10405/WBE-ethical-guidelines-v1.0–03.2016%20.pdf (accessed 2 December 2017).

Prichard, J., F. Y. Lai, J. O'Brien et al. 2018. 'Ice rushes', data shadows and methylamphetamine use in rural towns: Wastewater analysis. *Current Issues in Criminal Justice* 29:195–208.

Rigg, K. K., and S. M. Monnat. 2015. Urban vs. rural differences in prescription opioid misuse among adults in the United States: Informing region specific drug policies and interventions. *International Journal of Drug Policy* 26:484–91.

Roche, A., and A. McEntee. 2017. Ice and the outback: Patterns and prevalence of methamphetamine use in rural Australia. *The Australian Journal of Rural Health* 25:200–9.

Roxburgh, A., W. D. Hall, T. Dobbins et al. 2017. Trends in heroin and pharmaceutical opioid overdose deaths in Australia. *Drug and Alcohol Dependence* 179:291–8.

Rural Health Information Hub. 2018. Substance Abuse in Rural Areas. www.ruralhealthinfo.org/topics/substance-abuse#effects (accessed 22 May 2018).

Sherman, L. W., H. Strang, G. Barnes et al. 2015. Twelve experiments in restorative justice: The Jerry Lee program of randomized trials of restorative justice conferences. *Journal of Experimental Criminology* 11:501–40.

Smith, K. B., J. S. Humphreys, and M. G. A. Wilson. 2008. Addressing the health disadvantage of rural populations: How does epidemiological evidence inform rural health policies and research?. *The Australian Journal of Rural Health* 16:56–66.

Spooner, C., J. Bishop, and J. Parr. 1997. Research methods for studying injecting drug users in a rural centre. *Drug and Alcohol Review* 16:349–55.

Spooner, C., and K. Hetherington. 2004. *Social determinants of drug use.* Sydney: National Drug and Alcohol Research Centre, University of New South Wales.

Sterling, R., J. Simicevic, E. Allouche, W. Condit, and L. Wang. 2010. *State of technology for rehabilitation of wastewater collection systems.* https://nepis.epa.gov/Exe/ZyPDF.cgi/P1008C45.PDF?Dockey=P1008C45.PDF

Substance Abuse and Mental Health Services Administration. 2017a. *Key substance use and mental health indicators in the United States: Results from the 2016 National Survey on Drug Use and Health* (HHS Publication No. SMA 17–5044, NSDUH Series H-52). Center for Behavioral Health Statistics and Quality, Substance Abuse and Mental Health Services Administration. www.samhsa.gov/data/sites/default/files/NSDUH-FFR1-2016/NSDUH-FFR1-2016.htm.

Substance Abuse and Mental Health Services Administration. 2017b. Results from the 2016 National Survey on Drug Use and Health: Detailed Tables. www.samhsa.gov/data/sites/default/files/NSDUH-DetTabs-2016/NSDUH-DetTabs-2016.pdf (accessed 1 June 2018).

United Nations Office on Drugs and Crime. 2017b. *World drug report 2017 (Booklet 4 – Market analysis of synthetic drugs: Amphetamine-type stimulants, new psychoactive substances).* New York: United Nations.

United States Department of Agriculture Economic Research Service. 2018a. Rural Classifications: Overview. www.ers.usda.gov/topics/rural-economy-population/rural-classifications/ (accessed 20 April 2018).

United States Department of Agriculture Economic Research Service. 2018b. Natural Amenities Scale. www.ers.usda.gov/data-products/natural-amenities-scale/ (accessed 20 April 2018).

United States Environmental Protection Agency. 2016a. *Clean Watersheds Needs Survey 2012: Report to Congress*. www.epa.gov/sites/production/files/2015-12/documents/cwns_2012_report_to_congress-508-opt.pdf.

United States Environmental Protection Agency. 2016b. *Clean Watersheds Needs Survey 2012 – Texas*. www.epa.gov/sites/production/files/2015-10/documents/cwns_fs-tx.pdf

Uporova, J., A. Karlsson, R. Sutherland, and L. Burns. 2018. *Australian trends in ecstasy and related drug markets 2017. Findings from the Ecstasy and Related Drugs Reporting System (EDRS)*. Sydney: National Drug and Alcohol Research Centre, University of New South Wales.

Wilkes, L. 1999. Metropolitan researchers undertaking rural research: Benefits and pitfalls. *The Australian Journal of Rural Health* 7:181–5.

chapter five

Micro Applications of Wastewater Analysis

Prisons, Educational Institutions and Workplaces

INTRODUCTION

Chapter 3 examined the utility of WWA in the global context in evaluating the effectiveness of international policy platforms. The book narrowed its scope in Chapter 4 to focus upon ways in which WWA might be of benefit within a country, particularly those with large geographical boundaries. In this fifth chapter we narrow our geographical scale further to explore micro applications of WWA, namely its possible use in specific buildings and clusters of buildings.

The civic value of studying substance use behaviours within particular buildings or clusters may seem weak in comparison to the large and complex challenges faced at the national, regional and international levels. Nonetheless, the substance use of groups of people within these settings can be extremely important. In our view this is most apparent in prisons. Demand for psychoactive substances within prisons can be very high because of the concentration of inmates who have previously used psychoactive substances, or who have current substance use disorders. There may be no shortage of people who are willing to profit from meeting that demand despite the risks and consequences of doing so. This combination of factors means that it is very difficult to keep psychoactive substances out of prisons. It is well accepted that substance use in prisons can cause an assortment of harms to inmates (e.g. overdoses, infections, assaults) and the broader community after their release (e.g. infection spread).

For these reasons prisons are the main focus of this chapter. We more briefly consider the possible use of WWA in educational settings and its potential limited application in certain workplaces. On prisons, we have delimited our discussion to the incarceration of adults for breaches of criminal laws. Other incarcerated populations, such as minors and refugees, are beyond our scope. We also have primarily concentrated on prisoners' consumption of illicit drugs and the diversion of pharmaceuticals.

The aims of this chapter are important to stress. It is not to build a case for prison WWA studies. Rather it is to scrutinise the potential risks, benefits and uncertainties of using WWA in prisons where small populations contribute to wastewater samples. There are substantial risks of harm, especially for the wellbeing of inmates. Just how great these risks are will vary significantly between countries and even individual prisons. However, there are also substantial potential benefits for inmates from WWA. Research teams backed by the expertise of registered human research ethics committees (HREC) can take responsibility for assessing where the balance lies on a case-by-case basis. Our view is that WWA may be able to inform and thereby improve policy. Among other things, this may be prevented where corrective services agencies are reluctant to facilitate evaluative research using any method (Rodas, Bode and Dolan 2012), or are uninterested in policy change (e.g. see McConnell and 't Hart 2019).

5.1 Overview of Prison Infrastructure

Prisons are a common feature of criminal justice systems around the world (UNODC 2019). Globally an estimated 10.74 million people were imprisoned in 2018 – either awaiting trial or post-conviction (Walmsley 2018). The true figure is probably higher because among other things the reporting of data is incomplete in some countries, such as China, and in other countries there are no data (Walmsley 2018, 2). Substantial resources are required to build and maintain prison buildings, care for inmates, sustain prison administrations and so forth. For example, analyses of public expenditure for 2010 suggest that the United Kingdom spent 0.37% of its GDP (gross domestic product) on prisons while the Netherlands spends 0.42% (EMCDDA 2014).

The world's prison population is increasing. But substantial differences exist between regions in the rate at which people are imprisoned. Some regions, such as the Americas, have experienced substantial growth in prison numbers for many decades (Csete et al. 2016). Figure 5.1 presents estimates of the numbers of prisoners per 100,000 people in each region of the world.

Walmsley (2018) estimated that the global imprisonment rate is 157 per 100,000 people. Figure 5.1 shows that regional rates vary from 97 per 100,000 people in Africa and Oceania, to 376 in the Americas.

There is also considerable variation in rates of imprisonment within regions. For instance, imprisonment rates in Africa range from 16 in the Central African Republic to 464 in Rwanda (Walmsley 2018). In the Americas, rates vary from 96 in Haiti to 655 in the USA.

Similar patterns are found between jurisdictions within national borders. By way of example, the state of Vermont in the USA had an estimated

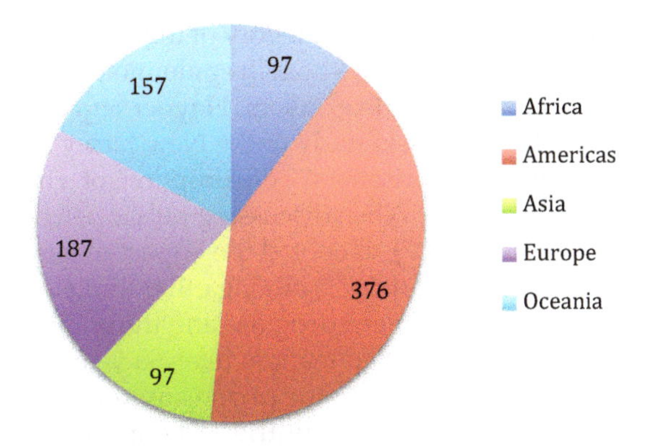

Figure 5.1 Numbers of Prisoners per 100,000 People by Region (2018 Estimates). Adapted from Walmsley (2018, 17).

imprisonment rate of 280 per 100,000 American residents in 2016 (Kaeble and Cowhig 2018) while in Mississippi in the same year the rate was 960. Racial minorities are overrepresented in the prison populations of countries such as the US, Australia, Canada and Brazil (see ABS 2018; Csete et al. 2016).

The incarceration of over 10 million people requires many prison complexes. Information about prison infrastructure is not readily accessible in many countries. Quite detailed statistics are available from the 2005 census of State and Federal correctional facilities in the US, which was published by the Department of Justice (Stephan 2008). Though dated, the census is useful in providing some metrics on one of the world's largest correctional systems, which incarcerated approximately 2.16 million people in 2016 (Kaeble and Cowhig 2018).

The census indicated that in 2005 the US had 1,292 prisons, 107 of which were privately operated.[1] The figure for 2005 represented an increase of 84 prisons since the 2000 census.

The features of the 1,292 prisons are not discernible from the report because it aggregates prisons with 'community-based' correctional facilities (N=529) (Stephan 2008). This category encapsulated institutions where more than half of the residents were regularly permitted to leave, sometimes unaccompanied. Examples of the community-based facilities included halfway houses and residential treatment centres.

5.2 *Prison Populations and Markets for Psychoactive Substances in Prison*

Given that this book is written for multiple disciplines, it is not our intention to outline the extensive literature that exists concerning the health of prison populations or penology. A large sub-category of the research on prisons has examined substance use by inmates – both in epidemiology (see Fazel, Bains and Doll 2006) and criminology (e.g. Akers, Hayner and Gruninger 1974). Sykes' (1958 (2007), 44, 97) classic study of the New Jersey State Maximum Security Prison referred to prisoners' consumption of contraband "fermented alcohol" and "hooch". When discussing public perception of prison management, Sykes (1958 (2007), 45) noted the impact of newspaper stories about prisoners' "illegal use of drugs", among other things.

Sykes' observations – prisoners' drug use and the willingness of media outlets to report it – are familiar to readers over 60 years later. However, substantial changes have occurred to the prison environment since the late 1950s in the US and in many other countries. The most obvious transformation has been a large increase in the number of prisoners incarcerated for drug-related offences. Sykes (1958 (2007)), for example, did not list drug offences among the main categories of convictions in the prison population he studied. By contrast, in many modern prison systems large proportions of inmates are incarcerated for drug offences. On 30 September 2016 approximately 47% of sentenced federal US prisoners were principally imprisoned for drug offences; i.e. their most serious offence leading to a conviction was a drug offence (Carson 2018). This equates to 81,900 people. For US state correctional authorities, Carson (2018) estimated that 15% (N=197,200) of inmates were primarily convicted because of drug offences (Carson 2018).

Varied statistics can be found in other countries. Approximately 16% of all Australian inmates were principally in custody for illicit drug offences as at 30 June 2018 (ABS 2019). In the UK 19% of remandees were held for drug offences at 30 September 2019 (Ministry of Justice 2019). Data collated by Csete et al. (2016, 1444) show that the proportion of inmates imprisoned for drug offences ranged from 10% in New Zealand to 68% in Thailand. Where there are data available about gender, globally women are over-represented in prisons for drug-related offences (Csete et al. 2016).

Drug laws vary internationally but there are two broad categories of offences: (a) those that relate to personal drug use and (b) those that relate to involvement in drug trafficking – ranging from minor street sales to high-level organised crime. This means that prisons contain people with an interest in using illicit drugs and people who have experience in drug trafficking. In economic terms, there is an interest in both the demand-side and the supply-side of the prison drug market.

5.2.1 Demand-Side Dynamics

Evidence from a variety of sources indicates that prisoners are much more likely than the general population to report having used an illicit drug before entering prison (e.g. AIHW 2019; EMCDDA 2012; Kevin 2013).

Substance use disorders are also more prevalent in prison populations than in the wider community (e.g. Bronson et al. 2017; UNODC 2019).

There are a variety of reasons why inmates may want to consume illicit drugs in prison, whether to initiate use, or to continue their substance use. Motivations for use may change across an inmate's time in prison and this, in turn, may influence the types of substances they consume (if available) (Cope 2000; Penfold, Turnbull and Webster 2005). Different studies have reported consumption by prisoners of major drugs of concern, such as cannabis, heroin, cocaine and amphetamine-type substances (UNODC 2019) – as well as alcohol (Mundt et al. 2018), new psychoactive substances (NPS) (HMI Prisons 2019; Ralphs et al. 2017) and diverted pharmaceuticals such as buprenorphine (van Dyken et al. 2016). Tobacco is also a contraband commodity in jurisdictions where smoking has been banned in prisons.

Among other factors identified in literature, prison drug use may be a means of: alleviating boredom (HMI Prisons 2019), reducing anxieties related to prison life and the loss of liberty (Boys et al. 2002; Cope 2000; Dean 2005; Loxley et al. 2004; Woodall 2012), and improving social relations with other inmates (Winfree, Newbold and Tubb III 2002; Woodall 2012).

It is important to note that prison may also provide some inmates with an opportunity to desist from problematic substance use (Woodall 2012). This may be because of the obvious changes in drug availability and their environment and routines (Crewe 2005; Woodall 2012) and because inmates may benefit from drug treatment programs that are not as readily accessible outside prison (Fazel, Bains and Doll 2006). As Crewe (2005, 474) observed, "while some prisoners find drugs a respite from prison, others find prison a respite from drugs."

Measuring the prevalence of drug use in prisons is difficult. Global estimates based on 149 studies in 62 countries suggest that 31% of prisoners have used illicit drugs in prison at least once and 19% have done so in the month prior to being surveyed (UNODC 2019). Plourde et al.'s (2012) study of Canadian male inmates showed that one-third had consumed illicit drugs while incarcerated. Similar rates have been reported in Australian studies (e.g. Kevin 2013; Richters et al. 2008). Estimates at US prisons have ranged between 22% and 60% (see Gillespie 2005, 225).

In the most recent survey of English and Welsh inmates conducted by HM Inspectorate of Prisons (2019), 20% of white men (N=4,031) and 10% of black and minority ethnic men (N=1,814) reported that they had developed an illicit drug problem while they were in prison. Slightly lower percentages reported that they had developed problems with diverted pharmaceuticals (11% and 7%, respectively). Of those who reported substance use in prison, half of the white men and one-third of the black and minority ethnic men indicated that it was "very/quite easy to get illicit drugs" in prison (HMI Prisons 2019, 100).

Drug markets within prisons can change over time (Plourde et al. 2012). Consumption levels can vary significantly between prisons at the national (Fazel, Bains and Doll 2006; Vandam 2009) and international level (Csete et al. 2016; EMCDDA 2012), including low- and middle-income countries (Mundt et al. 2018). By way of example, estimates of the numbers of inmates who have ever used illicit drugs while in prison are 7–53% in Mexico, 0–23% in Lithuania, and 2–5% in Romania (Mundt et al. 2018). Variation also has been observed between countries in Western Europe, although the comparability of data is limited because of differences in the survey methods employed by different research teams (EMCDDA 2012).

5.2.2 Harms of Prison Drug Use

There are cogent arguments why persons convicted for minor drug crimes should not be imprisoned. In addition to the disproportionality of imprisonment for personal drug use, incarceration can *itself be harmful* to offenders and their families and wider communities (Csete et al. 2016). Clearly, the effects of prison drug markets can exacerbate the negative effects of incarceration. Among other things, inmates who use illicit drugs can seriously endanger their physical and mental health. Prison drug use is associated with:

- inmate deaths from overdoses (Albizu-García et al. 2009; Dolan et al. 2007; Nunn et al. 2009);
- the transmission of blood-borne viruses (BBV), such as HIV and hepatitis C, particularly through the sharing of injecting equipment (ANCD 2008; Prendergast et al. 2004; Wood, Montaner and Kerr 2005);
- the exacerbation of mental illnesses and social problems (ANCD 2008); and
- sexual assault and unsafe sex practices (Kang et al. 2005; Steels and Goulding 2009).

The impact of substance use in prison is not confined to drug users. In fact, the ripple effect of drug use can be profoundly negative for other inmates and prison staff. It is well known that prison drug markets can increase levels of violence within prisons (Prendergast et al. 2004). For instance, drug sellers may use violence and intimidation to enforce payment of drug debts (Kevin 2013). Assault can be used to rob inmates of prescription medications (Penfold, Turnbull and Webster 2005). High frequency substance users can be hostile to other prisoners when intoxicated and especially when suffering symptoms of withdrawal (MacPherson 2004).

For corrective services staff in some prisons "drug-fuelled violence remains a daily reality" (HMI Prisons 2019, 7). The recent official description of one UK prison underscored how the negative effects of drug consumption can have a paradoxical effect of increasing demand for drugs.

> Onley was a clear example where the failure to deal with drugs and violence undermined many other aspects of prison life… fear, frustration and boredom increased the demand for drugs, which in turn fuelled the violence, and thus completed the circle.
>
> (HMI Prisons 2019, 23)

While the jurisprudence rationale for imprisonment varies internationally, many prison systems do attempt to rehabilitate inmates and maximise their chances of reintegration back into society. There is no doubt that drug markets can undermine rehabilitative programs within prisons (Dolan et al. 2007). Reintegration is also affected; post-release, former inmates with substance use disorders are at higher risk of relationship breakdown, unemployment and recidivism (ANCD 2008; MacPherson 2004). Among other implications for society, the use of drugs in prison has been shown to increase rates of blood-borne diseases in the population (Levy et al. 2007).

5.2.3 *Supply-Side Dynamics*

It is very difficult to prevent all illicit drugs from entering prisons. Given that the value of illicit drugs may increase three or fourfold within prison walls (Crewe 2005), traffickers use inventive strategies to smuggle substances past prison security systems. For example, drones were employed to fly illicit drugs reportedly worth £500,000 into UK prisons on 55 occasions between 2016 and 2017 (BBC News 2018). Drugs can be delivered through the prison mail system (CCC 2018), or carried into the prison within the body cavities of inmates or visitors (Ralphs et al. 2017). In Australia it was alleged that ex-inmates deliberately breached their parole conditions so that they would return to prison and thereby have the opportunity to traffic drugs internally (Jacques 2019). Pharmaceuticals medically administered within the prison can be diverted at the point they are dispensed (Penfold, Turnbull and Webster 2005).

An important variable affecting prison drug markets is the potential for corruption among corrective services staff. This includes not only guards, but other employees such as instructors, health workers and allied health professionals (CCC 2018; Souryal 2009). Significant numbers of people are employed in some countries by corrective services. In the US, the Department of Justice estimated that across all prisons and community-based institutions approximately 445,000 people were employed (Stephan 2008). Two-thirds of the employees (N=295,261), including correctional officers, "worked in direct contact with inmates and were involved in their daily custody or monitoring" (Stephan 2008, 4).

Corruption within prison systems is a complex topic of international interest (e.g. UNODC 2015; 2017). Corruption varies in term of its seriousness, ranging from ignoring minor breaches of regulation at one end of the spectrum, to involvement in financially motivated trafficking at the other (CCC 2018). Staff who engage in low-level drug trafficking may later be blackmailed by inmates into participating in larger deals (Crewe 2005). Corruption that is detected by law-abiding staff may not be reported because of subcultural norms about the need for secrecy and the importance of protecting colleagues (CCC 2018). In contexts where corruption is widespread, previously law-abiding staff may be pressured to participate in the activity to reduce the risk of reporting (UNODC 2017).

5.3 Strategies to Respond to Substance Use in Prison

Harm reduction strategies in prisons aim to reduce the prevalence of drug-related harms, without necessarily reducing drug use. For example, needle and syringe programs (NSP) are designed to reduce the spread of BBV (Csete et al. 2016). Other examples of harm reduction strategies that lower risks of BBV and overdose (Marlowe 2011) are opioid substitution therapy (OST) (Larney et al. 2014) and behavioural interventions to reduce substance misuse. They can also be defined as *demand reduction strategies* because they enable inmates to reduce their use of illicit drugs.

Prison security systems that are designed to prevent drugs and other contraband from entering prisons can be described as *supply reduction strategies*. These include features of prison architecture, such as walls, perimeter fences, surveillance posts and cameras, and security gates. Some prisons are resourced with a range of sophisticated equipment, such as telephone-blocking technology (to prevent communications on trafficking), drug-detection kits for prison mail and X-ray equipment to scan bodies and baggage (Ministry of Justice 2019).

Cell and visitor searches, drug-sniffer dogs, strip searches and perimeter patrols are commonly used to locate drugs to reduce supply. In principle, they also reduce demand by acting as a deterrent to possessing drugs (e.g. ANCD 2008; Hughes 2000). Mandatory drug testing (MDT) using urine is employed in many prisons. When there are penalties for positive tests (e.g. Prichard et al. 2010), MDT can be categorised as a demand reduction strategy.

The use and quality of these strategies clearly differs between prisons (Csete et al. 2016), even within jurisdictions. National assessments in Australia have raised concerns about an over-reliance on supply reduction strategies (Black, Dolan and Wodak 2004; Rodas, Bode and Dolan 2012). Recent observations in parts of the UK by HM Inspectorate of Prisons (2019) found that the types of technology used to reduce supply varied and that there was insufficient delivery of demand reduction strategies – with only half of those inmates who reported a drug problem receiving help in prison (HMI Prisons 2019, 199). Finally, only a small number of prisons ensured that demand and supply reduction strategies were integrated and appropriately coordinated.

5.4 *Wastewater Analysis in Prisons*

Wastewater analysis has been used to examine substance use in prisons in Spain (Postigo, López de Alda and Barceló 2010), the USA (Brewer et al. 2016), France (Néfau et al. 2017) and Australia (van Dyken et al. 2014, 2016). It is unclear whether WWA researchers have also conducted confidential fee for service analyses of wastewater for corrective services. Below we consider the potential limitations and strengths of WWA for prison research.

The section on scientific limitations (5.4.1) briefly broadens our scope beyond prisons because of its relevance to the use of WWA in sites such as workplaces and schools. Understanding how WWA can function in prisons is important for gauging its value as a research method and for critically examining ethical issues in research. Researchers from any discipline considering a WWA-prison study need to carefully consider the ethical context because it informs fundamental questions such as should the study be conducted and, if so, how can any ethical risks be mitigated?

5.4.1 *Building Complexes (Including Prisons): Specific Issues Relating to Sampling and Laboratory Analyses*

In Chapter 2 (2.2.2) we explained that the validity and quality of WWA data depends on the procedures used to collect and store samples. In fact, more error can be introduced into WWA data through collection and storage than by the laboratory techniques used to analyse the samples (Ort et al. 2010). In collection the key objective is to employ a procedure that maximises the degree to which the wastewater samples are representative of all the wastewater that has flowed into a treatment plant over a given time period (e.g. 24 hours).

Flow is not constant in large sewers systems. It is constituted by irregularly discharged 'packets' of wastewater. Variation in flow can be even *greater in small sewer systems*, such as those servicing a prison complex, because of the shortness of the sewers and the lack of retention tanks in gravity sewers (Prichard et al. 2010).

Variation in concentration may be significant as well. For instance, some water packets might contain comparatively high quantities of urine (e.g. when inmates' morning routine commences) and others low (e.g. when prison laundry is cleaned and greywater output increases). Because of the assumptions made in the laboratory when the 'load' (weight) of different substances is estimated (see 2.2.3), failure to account for sampling factors may lead to gross underestimation or overestimation of psychoactive substance use in a building complex.

From a sampling perspective, the quality of sewer infrastructure servicing specific sites varies considerably. Many site-specific WWA studies have been conducted without encountering complexities with sampling (e.g. Gushgari et al. 2018; Lai et al. 2013). But WWA may not be feasible at some sites because of the design or functioning of the sewer system. The Australian prison our team studied (van Dyken et al. 2014; 2016) presented many challenges. Because of the degree to which the sewers leaked and other factors, separate analyses were conducted to estimate how much wastewater was lost by the time it reached the collection point for the auto-sampler. Since the system did not have its own retention tank or weir, estimating flow was time consuming. Additionally, the mechanism of the auto-sampler was placed under strain because it sat atop a manhole approximately 1.7 m above the sewer – as opposed to a much shorter distance than might be expected next to a weir.

The data we collected were scientifically valid. However, to account for uncertainties we needed to gather a larger number of samples than originally anticipated. This underscores an important point. The estimated validity of any one data point (e.g. a single sample) will depend on the quality of the sampling conditions. At one end of the spectrum, an optimal sampling site will produce high quality data meaning that conclusions could confidently be drawn with fewer samples than from a site at the other end of the spectrum which offers sub-optimal sampling.

In addition to examining sewer infrastructure, it is important to understand how a building complex is used and what groups constitute the target population for the research questions. Sewer infrastructure may enable sampling that relates exclusively to the relevant population. For example, prison sewers may permit sampling of wastewater produced exclusively by inmates. In other cases the samples will contain wastewater produced by inmates, prison staff and visitors. The contribution of visitors was complicated for Brewer et al. (2016) to control for because of the length of daily visiting times (6 hours and 15 minutes). However, in our study staff contributions were minimal (most were serviced by a separate sewer) and we were able to adjust the sampling regime around the short daily visiting times (<1 hr) to minimise input from visitors.

One of the points made in Chapter 2 was that although care must be taken in setting up sampling systems, with basic training personnel from collaborating agencies (e.g. water authorities) can follow appropriate procedures for collecting samples, preserving them and transporting them to laboratories, which in some cases can be thousands of kilometres from the site of collection. The same applies to site-specific WWA studies. For instance, the site of our prison study was over 1,700 km away from the laboratory where the analyses were conducted.

The limitations discussed above all relate to *sampling* from building complexes. A different type of scientific limitation needs to be reiterated here that relates to conducting WWA studies of small populations. This issue was discussed in 2.2. As explained, the *back-calculation process* is based on estimates of population. It may be possible for WWA researchers to have access to metrics about the population of a building complex that are much more precise than would be the case for a study of a suburb or city (e.g. van Dyken et al. 2014; 2016). Prison authorities can also provide reliable data about the types and quantities of medications that are dispensed by health professionals to inmates. These data may be important for the back-calculation process so that therapeutic use of a drug, such as morphine, is not erroneously counted as illicit use of a substance such as heroin.

However, back-calculation also draws on information about the average human metabolic processing of each substance (e.g. types of biomarkers excreted, their ratio to each other, and the average rate of excretion). It is legitimate to apply these averages in large populations because it can be expected that individual metabolic outliers – whose rates can differ significantly from other members of the population – will be relatively few in number. But this does not hold true in small populations; fluctuations may be due to which individuals are using a substance rather than the amount of the substance is being consumed. This means that, though site-specific WWA studies can produce valuable data, additional caution is required when interpreting findings – particularly in attempts to estimate the numbers of doses consumed (e.g. van Dyken et al. 2014; 2016).

5.4.2 Implications for Project Management

What are the implications of the issues raised here for agencies and research teams interested in a WWA for a building complex? Four main points should be considered for project management:

1. Early discussions with stakeholders (e.g. corrective service agencies) need to raise the importance of a feasibility study at the building or buildings in the proposed research.
2. Feasibility will be determined by sewerage engineers (e.g. by visiting the site and learning about how the sewers operate) and by analytical chemists drawing on information about population size, population movement and so forth.

3. Sub-optimal sampling conditions will increase costs associated with, for example, conducting on-site tests of the sewer environment and renting sampling equipment.
4. Optimal sampling conditions introduce fewer uncertainties than sub-optimal conditions and, as noted, conclusions can be drawn with *fewer samples*. Because optimal sites provide greater sensitivity and require shorter data collection periods, they provide more options in terms of research designs (see 1.3.2) – such as measuring whether a change in policy appears to be related to a change in drug consumption, discussed further below.

5.4.3 When Scientifically Valid, What is the Potential Utility of WWA in Prisons?

Wastewater analysis in prisons would not be worth considering if it could be stated that prison drug use is consistently: (a) not very harmful, (b) easy to prevent and (c) simple to measure. However, this is not the case. As we have seen, preventing the harms of drug use in prison remains a major problem for governments worldwide.

With regards to measurement, some scholars have recognised the need to improve the means by which inmates' drug use is quantified so that the effectiveness of prison strategies can be evaluated (e.g. Black, Dolan and Wodak 2004; Rodas, Bode and Dolan 2012; Royuela et al. 2014). A wide variety of data can be drawn upon to give a picture of the extent of consumption in prisons (e.g. data relating to drug seizures, violence, weapon seizures and demand for treatment services) (see e.g. Prendergast et al. 2004). When combined with survey data, analyses of prisoners' hair may be useful (Harrison 1997; Shearer et al. 2006). Ethnographic studies provide rich qualitative data about drug use and the lived experience of prisoners (e.g. Crewe 2005) but limited information on the prevalence and frequency of use.

The two most commonly used methods for measuring prison drug use are *surveys* and *MDT*. Consistent with our recommendations in earlier chapters, we think the value of WWA lies in supplementing rather than replacing these methods. A key question then is: does WWA provide valuable information that surveys and MDT cannot?

5.4.3.1 Prison Surveys

The advantages that surveys have in the general population (see 2.5) also apply to prison populations. Unlike WWA, surveys can provide information on individual inmates' substance use such as:

- the types and purity of drugs used;
- the frequency of use;

- whether they engage in polydrug use;
- perceived effects of drug use on inmates' health;
- risk behaviours related to routes of administration (e.g. needle sharing);
- the cost and means of payment for drugs;
- associations between drug use and violence;
- inmates' perceptions of the effectiveness of strategies to reduce supply, demand and harm;
- risk factors for substance use over the life-course.

Survey methods are not affected by the prison sewer infrastructure, as is the case for WWA. And the social science skills required to conduct robust prison surveys are easier to source than the technical knowledge required for valid WWA research.

In general, survey participants are less inclined to provide accurate information if they fear negative consequences from doing so (Harrison 1997). So it seems feasible that, in some consequences at least, prisoners may underreport their use of drugs in prison (Hughes 2000; Prichard et al. 2010). This may be due to their fear of traffickers – who have a vested interest in discouraging outside interest in prison drug markets – more than any fear of prison authorities.

Prisoners have been found to provide accurate information in surveys about their health (e.g. Schofield et al. 2011). McGregor and Makkai (2003) found good agreement between the drugs found in urine samples provided by arrestees and their self-reported recent drug use. Researchers can also adopt procedures to increase inmates' willingness to honestly report their substance use while in prison, such as recruiting inmates who are nearing the end of their sentence and using skilled interviewers (Kevin 2013). For these reasons surveys can be employed not only in one-off studies, but as an ongoing monitoring tool to develop trend data (Kevin 2013). Perhaps the best publicly available example of this currently is the annual survey of thousands of inmates conducted within the UK (HMI Prisons 2019).

However, there is no doubt that prison surveys are more complicated to conduct than those in the general population, as evidenced by the fact that regular prison surveys are uncommon (Royuela et al. 2014). Prison populations tend to have low levels of education so skilled interviewers are needed to build rapport (Kevin 2013) during face-to-face interviews to counter problems with literacy (Royuela et al. 2014). Corrective service agencies may be unwilling to permit research because of the in-kind costs associated with facilitating interviews and safeguarding interviewers' safety while in prison. Even if official approval is provided by prison authorities, correctional staff may not actively support survey research and thereby reduce response rates (Gillespie 2005). Like other populations who are of special interest to researchers from multiple disciplines, prisoners can experience research fatigue and may be disinclined to participate in surveys. These and other factors can contribute to non-representative samples of prison populations (Gillespie 2005; Royuela et al. 2014).

Prison surveys do yield details about the types of psychoactive substances inmates have consumed, including tobacco, alcohol, NPS and diverted pharmaceuticals (Royuela et al. 2014). However, it is difficult for prison surveys to provide detailed temporal data on drug use patterns. For instance, the European surveys reviewed by Royuela et al. (2014) asked inmates about their substance use in the past year and past month.

5.4.3.2 Mandatory Drug Testing (MDT)

Urine tests provide biological indicators of individual inmate's recent substance use. Like survey methods, MDT has a wider applicability than WWA because it is not limited by the technical requirements for sampling. Within many prisons systems inmates can be legally compelled to supply urine samples; inmates who refuse to supply samples can be punished as can those whose samples indicate that they have used an illicit drug while in prison (e.g. Prichard et al. 2010). MDT can be used by prison authorities to target inmates suspected of drug use. MDT can also be conducted randomly. In many contexts the main objective of urinalysis is to deter drug consumption by prisoners, and inmates have sometimes reported that urinalysis has a greater deterrence effect than other measures (Kevin 2013; cf Black, Dolan and Wodak 2004; Brewer et al. 2016).

From a research perspective, random MDT has greater potential to generate statistics that are broadly representative of the prison population as a whole. By contrast, targeted MDT results are difficult to interpret. If the number of positive results from targeted MDT increases or decreases, it cannot be determined if this is due to (a) fluctuations in drug consumption by prisoners or (b) changes in the effectiveness of prison staff in accurately targeting inmates who have used drugs.

However, the resources required to conduct random MDT on a large scale can be expected to be significant, given that in one year two Australian states respectively conducted 30,718 and 11,130 urine tests on prisoners (Black, Dolan and Wodak 2004). Consequently, scholars in some jurisdictions have recommended discontinuing random MDT in favour of targeted MDT (Dolan and Rodas 2014; see also Gore and Bird 1996). According to MacPherson (2004), the enforcement of random MDT can generate tensions between staff and inmates because of the adverse consequences for prisoners whose tests are found to be positive for drug biomarkers.

In addition, there are a number of reasons why random MDT may not provide an accurate snapshot of drug consumption across a prison. If random MDT is conducted on a regular schedule, such as certain days of the week (e.g. Hughes 2000), then inmates can use drugs when they know that the risks of detection are lower. It is also likely – given the forms of staff corruption discussed earlier (CCC 2018; UNODC 2017) – that some prison staff may assist inmates to avoid detection, for example by giving advance notice of MDT. MacPherson (2004) and Hughes (2000) highlighted ways in which inmates can undermine the accuracy of MDT, such as directly adding soap or vinegar to their sample, or drinking large volumes of water to dilute the concentration of biomarkers in their urine. Prison authorities have also told us that under regulations in their jurisdiction, inmates can technically comply with MDT laws by providing a very small amount of urine that is insufficient for analysis.

Other issues relate to the reliability of urinalysis. Illicit drugs are metabolised at different rates so MDT can produce false negatives and false positives (Dean 2005; Kendall and Pearce 2000; MacPherson 2004). For instance, recent use of a substance like methamphetamine or heroin might not be detected because it typically passes through the body within 24 hours (false negative). Conversely, MDT may erroneously suggest that an inmate's use of cannabis was recent, because biomarkers for cannabis use can take many days to leave the body (MacPherson 2004). Some scholars have suggested that this feature of MDT may inadvertently encourage inmates to use 'harder' drugs in preference to 'softer' drugs, such as cannabis (e.g. Hughes 2000; Kendall and Pearce 2000). However, others have questioned this claim (Crewe 2005).

5.4.3.3 Comparison of WWA with Surveys and MDT

The main empirical strengths of WWA in the prison context are summarised below. Given that only four prison studies have been conducted worldwide, these are still *potential* benefits that do not have as good an evidence base as WWA in the general population.

5.4.3.3.1 Impost on Prisoners and Staff First, WWA can be conducted with minimal impost on staff and no impost on inmates because it can be conducted outside the walls of the prison complex. This means that it avoids many of the problems that surveys encounter with respect to: the burden of in-kind costs for corrective service agencies; uncooperative correctional staff; or research fatigue for staff and inmates. Similarly, unlike random MDT WWA is unlikely to trigger tense interactions between staff and inmates because it does not interfere with day to day life within the prison.

5.4.3.3.2 Visibility The visibility of data collection for WWA depends on the sewer infrastructure and the location of the sampling point. The least we would expect is that sampling can be conducted discreetly. The main advantage of this aspect of the method is that it would be difficult for inmates to influence results by reducing their drug use when WWA sampling is being done – which is one of the potential problems facing MDT on a routine schedule. In situations where prison authorities are concerned about staff corruption, WWA may be conducted covertly without the risk of staff informing inmates of data collection.

5.4.3.3.3 Data Quality for Monitoring and Intervention Studies As noted, WWA cannot provide the person-centred drug use data that surveys yield. Nor is it designed to have, like MDT, a deterrent effect by detecting recent drug use by individual inmates. However, as a method to monitor drug consumption across a prison population, WWA can have significant advantages when sewer infrastructure enables optimal sampling conditions and we can estimate the amount of wastewater contributed by non-inmates.

WWA can provide data on a wide variety of drugs consumed by an entire prison population every 24 hours. For drug monitoring this arguably makes WWA superior to MDT, which might fail to detect use of certain drugs because too few urine tests are conducted (as found by van Dyken et al. (2016) on the non-therapeutic consumption of buprenorphine). MDT may also fail to detect drug consumption because of the time lag between inmates' use and the time of testing. For similar reasons, the monitoring capacity of WWA is better than prison surveys, the value of which can be undermined by long data collection periods, small sample sizes, selection bias and lack of temporal clarity (i.e. at what date inmates used drugs).

As we have proposed elsewhere (Prichard et al. 2010), in optimal sampling conditions WWA may be of sufficient quality to be used in a variety of empirical research designs. The designs could move beyond monitoring trends in the consumption of different types of illicit drugs (and alcohol or tobacco) to examining the relationships between prison drug markets and other variables, including prison supply, demand and harm reduction strategies. The exact number of data points – that is, days of sampling – required for statistical power and confidence would depend on the complexity of the research questions and the types of analyses conducted.

WWA could be used to conduct **natural experiments** (see further 1.3.2). As a simple example it could evaluate the impact of prison authorities building an additional fence around the perimeter of a prison to reduce the supply of drugs and other contraband. A WWA study could monitor the impact of this supply reduction strategy on prison drug market pre- (period A) and post-construction (period B) by sampling at monthly intervals for a year. Among many other possible findings, the study could indicate that the fence:

- had no impact on drug consumption;
- reduced consumption of some drugs while increasing the use of others;

- reduced total drug consumption for the first two months, but the effect dissipated over 12-months as alternative methods of drug supply developed.

Any of these findings would be useful for the managers of the prison at the focus of the study and for corrective service agencies more broadly.

Demand reduction strategies are likely to be more difficult to study with this sort of A-B design in most cases. This is mainly because (a) it requires sustained efforts over time to change inmates' drug use and (b) many prisons have a steady turnover of inmates, which means a constant flow of new inmates with problematic substance use.

There are feasible exceptions. For example, if corrective service agencies decided to *implement* (i.e. start) a comprehensive OST program (e.g. with buprenorphine) to reduce demand for methamphetamine (MA), then a WWA-based natural experiment could be conducted. Period A (baseline) would monitor MA consumption before OST was introduced. Period B could do two things. First, it could assess whether the OST reduced MA consumption, and if so for how long. Second, it could compare WWA data with metrics provided by the prison indicating how many doses of buprenorphine were administered each day, as demonstrated by van Dyken et al. (2016). The latter approach could be used to detect whether OST had increased the non-therapeutic use of buprenorphine.

Harm reduction strategies are not designed to influence drug consumption levels but nonetheless a relevant question for researchers, agencies and policy makers may be whether they increase prison drug use. A good example would be NSP. Despite evidence that NSP are not associated with increased prison drug use (e.g. Stöver and Nelles 2003), the concern is still raised whenever NSPs are proposed in prisons (e.g. Bresnan 2015). A WWA study could play an important role assessing the impact of NSP on the prison drug market.

Additional steps could be taken to empirically strengthen the design of all these evaluations. Large jurisdictions could use two or more prisons with optimal sampling conditions (cf Néfau et al. 2017), as **comparator prisons**. These could act as controls for the prison which is the site of the natural experiment. Monitoring multiple prisons simultaneously in periods A and B would control for extraneous factors, such as fluctuations in the drug market outside the prison walls so reducing the risk that changes in inmates' drug use were erroneously attributed to a new supply, demand or harm reduction strategy.

In large prisons with sewer systems that allow optimal sampling at different prison blocks, then **comparator prison blocks** could be used as controls. In this scenario, a new supply, demand or harm reduction strategy would be implemented in the experimental block but not in the control blocks. This approach would control for the effects of any extraneous variables that are specific to an individual prison. For example, if during the B phase (while the strategy was implemented) there was an attempted escape the entire prison might be locked down for a period and drug supply within the prison disrupted. A comparison of drug use between the experimental and control blocks would enable researchers to assess the effect of the prison lock down on the WWA data.

Confidence in evaluations would be increased if studies were able to employ **withdrawal and reversal designs** (see Gast, Ledford and Severini 2018). Some interventions are limited to A-B designs because the policy change cannot be undone; e.g. the addition of a new perimeter fence. However, in other situations prison authorities may be able to implement *and withdraw* a strategy relatively quickly. This already occurs with some of the deterrent strategies discussed earlier, such as targeted MDT, cell searches, sniffer dogs and so forth. Where prison authorities collaborated closely with WWA researchers, A-B-A-B designs might be considered. This would be useful if, for instance, authorities wanted robust evidence of the efficacy of MDT as a way of reducing drug use. The A phase would involve gathering baseline data on the prison drug market. In the B phase MDT might be applied for a discrete period. After MDT is *withdrawn* (marking the end of the first A-B phase) the process would be repeated (introducing the second A-B phase). This design would be particularly useful in situations where control prisons or control blocks were absent. This is because data from the two A-B periods could increase confidence that any observed changes in the prison drug market were not due to extraneous variables internal to the prison (e.g. a lock-down) or external (e.g. fluctuations in the macro drug market). Several other forms of withdrawal and reversal designs could feasibly be explored with prison authorities to examine the effects of multiple strategies individually and in combination (e.g. A-B-C-B-C, A-B-BC-B-BC; see Gast, Ledford and Severini 2018, 225).

5.4.3.3.4 Cost Studies comparing the cost of prison surveys, MDT and WWA have not been done. We suspect that true costing (i.e. accounting for in-kind costs associated with staff time) would reveal that WWA is cheaper than large-scale random MDT. However, where resources are concerned, arguably the real potential benefit of WWA lies in the quality of its data – which enables it to robustly evaluate the impact of prison drug strategies including their unintended consequences and their cost-effectiveness (see Prichard et al. 2010).

In our experience one of the major factors influencing the cost of prison WWA research is the quality of the sampling. As noted (5.4.1), if the sampling conditions are sub-optimal additional resources may be required (a) at the commencement of the project to test the sewers and set up equipment and (b) during the data collection phase because more data points will be required to compensate for uncertainties. The number of samples required will be shorter for basic monitoring and longer for the more complex designs as discussed above. Our Australian study in sub-optimal sampling conditions involved basic monitoring and 24 days of sampling (van Dyken et al. 2016).

Given the importance of sampling conditions for the feasibility of WWA prison research, it is worth considering some implications for the design and construction of new prisons. As noted earlier, between 2000 and 2005, 84 new prisons were constructed in the US (Stephan 2008). While the architecture of prison buildings draws significant attention from various stakeholders and researchers (e.g. Hancock and Jewkes 2011), the design of sewer infrastructure has solely focussed on waste management. We recommend that agencies overseeing the design and construction of new prisons consider integrating features within the sewers that maximise wastewater sampling conditions – such as a weir or retention tank in the case of conventional gravity systems, a power source (for sampling equipment) and separate sewers for staff and visitors. While this may not be feasible in some circumstances, in others implementing these features would make a minimal contribution to overall construction costs.

5.4.4 Ethical Considerations

WWA researchers contemplating prison research are responsible for ensuring that they comply with the laws and regulations of their countries and the ethics requirements of their institutions. In this chapter we have focussed on prisons that hold adults who have breached criminal laws. It is essential to note that more weighty legal and ethical complexities may be encountered in studying drug use in other incarcerated populations, such as young people and people who are seeking asylum.

Internationally recognised ethical principles that guide human research were discussed in Chapter 2 (2.3). In précis, these require that the ethical merits of a proposed research project consider the degree to which it:

- respects participants' personal **autonomy** (e.g. demonstrated through informed consent and the maintenance of confidentiality and privacy);
- has reasonable prospects of benefiting the participants or broader society (**beneficence**);
- fairly and equitably distributes the burdens and benefits of research participation (**distributive justice**); and
- avoids or minimises harm to participants *and non-participants* (**non-maleficence**).

Prison populations are very small in comparison to the catchment populations in most WWA studies of the general population. Nonetheless, it would be impossible to identify the drug use of a particular inmate through WWA. This means that prima facie the principle of **autonomy** is not threatened despite the fact that inmates' data are analysed without their consent.[2] Equally **distributive justice** is satisfied in that prison WWA research places *no burden* on inmates and any benefits from the research are likely to advantage all inmates to some extent (e.g. through a reduction in drug-related violence).

The crux of the ethical assessment required for prison studies is the balancing of the principles of **beneficence** and **non-maleficence**. Clearly the *potential* benefits of a WWA study are very significant. They include: reductions in drug-related deaths and BBV; improvements in the prospects of inmates' reintegration after their release; increased physical safety of inmates and staff; improved inmates' psychological wellbeing; reduced strain on inmates' interpersonal relationships; and the more efficient use of prison resources.

But the risks of harm from WWA prison research are also very significant. It is for this reason that the WWA ethical guidelines recommended that researchers seek the approval of a human research ethics committee (HREC) before commencing a prison study (Prichard et al. 2016). In countries such as Australia it is likely that HREC oversight would be obligatory. Before considering types of harm, it is important to reiterate that for human research ethics the threshold for harm is low and may include distress, embarrassment or stigmatisation (NHMRC 2007 (updated 2018)) (see 2.3).

As a vulnerable group, inmates, ex-inmates and their families could easily feel distressed or embarrassed if the findings of a WWA study were reported in the media – particularly if the coverage named the specific prison in which they or their family member were incarcerated (Prichard et al. 2016). Misleading or sensationalised media reports could exacerbate inmates and ex-inmates' perceptions of stigma and devaluation by the community. These outcomes could reduce reintegration after release from prison because stigma predicts poorer outcomes for employment and community functioning in terms of residential stability, marital status, additional education and so forth (Moore, Stuewig and Tangney 2016; see also Winnick and Bodkin 2008). Feasibly, stigma may also manifest as discrimination within the community (e.g. Link and Phelan 2001, 371), for instance, towards ex-inmates from a prison that had been labelled as having a worse drug problem than other prisons.

The WWA ethical guidelines suggest strategies that may be used to mitigate these risks (Prichard et al. 2016) (see Chapter 2 (2.3)). They include anonymising the location of prisons in publications. Our experience, based on one study and two articles was that this measure effectively ensured the anonymity of the prison.

Another area of potential risk is the way in which WWA data might influence the policies of corrective service agencies. Institutions vary in the degree to which they pursue supply, demand and harm reduction strategies. It is the possibility of excessive, or "draconian" (Dean 2005, 168), supply reduction strategies that is most relevant to WWA. This is first because:

- WWA intervention studies most readily lend themselves to evaluations of supply reduction policies; and
- compared with other policies, supply reduction strategies are more likely to meet at least the lower threshold for harm as defined in research ethics – despite the fact that it is legal to implement them.

This last point is important. It shows that the boundaries of WWA researchers are different to those for the prison authorities with whom they may collaborate. The "legitimate" measures (Prichard et al. 2016, 8) open to authorities will depend on the legal framework governing their jurisdiction, which may include international human rights conventions, constitutional law, domestic legislation and common law.

However, if such measures could cause harm to participants or non-participants (e.g. visitors, staff) from the perspective of an ethics committee, then the principle of non-maleficence may not be satisfied. Establishing beneficence may also be difficult when prison authorities have a preference for strategies that lack a credible evidence base, e.g. cell searches or sniffer dogs. In such circumstances an ethics committee may decide that a project should not proceed or should be abandoned.

Globally the interpretation and application of the principles of human research ethics are far from uniform.[3] It is nonetheless useful to consider some examples of supply reduction strategies that arguably would be viewed as harmful by HRECs:

- the removal of contact visits – on the basis that it may negatively affect children's development and child-parent relationships (e.g. Poehlmann et al. 2010);
- invasive body searches that are likely to humiliate or distress inmates, adult and child visitors or prison staff;
- lock-downs that confine inmates to their cells for long periods of time; and
- the use of solitary confinement for inmates who fail MDTs.

How are researchers to assess risks associated with institutional responses to WWA data? The WWA ethics guidelines recommend that researchers seek to understand the ethical practices of prison authorities by taking practical steps, such as:

- reviewing the agency's drug policies;
- examining whether the agency has been criticised for violating inmates' rights, for instance by independent government bodies, courts or human-rights organisations (e.g. Metzner and Fellner 2010);
- liaising with researchers from other disciplines who can attest to the agency's ethical standards; and
- conducting preliminary meetings with the agency to assess these issues (Prichard et al. 2016).

Other points need to be considered here. First, it is important to appreciate the suitability of the individual prison or prisons for WWA research. For many reasons, practices and policies may differ markedly over time between prisons, even when they are governed by the same corrective services agency. The agency itself may have sound drug policies and a good record with regard to its care of inmates. However, its reputation may not apply to all prisons within that governance structure.

Second, most WWA researchers, including the authors of this book, operate within liberal democratic principles that encompass the rule of law, civil and political liberties, the separation of powers and so forth. Grave ethical risks arise if WWA prison studies are contemplated in countries that operate outside of these political constraints. For example, in our view it would be difficult to be confident that any of the principles that guide ethical conduct in human research would be satisfied in a country such as China which has demonstrated a disregard for inmates' rights (e.g. Paul et al. 2017).

It is possible that, despite researchers' best efforts to understand approaches taken by particular prisons, it is only after a project is under way that they discover that prison authorities intend to use WWA data to justify the use of strategies that may harm participants or non-participants. Suspending or abandoning the project may be necessary. This step will be less difficult if, as recommended by the WWA ethics guidelines (Prichard et al. 2016), researchers explain the project's ethical constraints to the prison authorities in early discussions of the study. In this situation, HREC oversight simplifies stakeholder management because it means that, among other things, (a) agencies cannot exert pressure on teams to alter their approach and (b) relationships are less likely to be strained because the decisions are made by an independent oversight body rather than by the research team.

5.5 Wastewater Analysis in Education Settings and Workplaces

WWA studies have been conducted in building complexes other than prisons, including an international airport (Bijlsma et al. 2012). In this final section we overview issues with respect to educational institutions and workplaces. The main technical limitations of WWA in these settings were discussed above (5.4.1). In our view, HREC oversight will be necessary in most of these site-specific studies.

5.5.1 *Educational Institutions*

Substance use is associated with, among other things, poor educational outcomes for young people (e.g. Prichard and Payne 2005). Estimates of the prevalence of substance use among students inform policy and practice. To date five WWA studies have examined substance use in American universities and colleges. These have provided useful data about the consumption of illicit drugs (Heuett et al. 2015; Gushgari et al. 2018; Panawennage et al. 2011), and non-therapeutic consumption of fentanyl and stimulants used to treat attention deficit hyperactivity disorder (Burgard et al. 2013; Gushgari et al. 2018; Moore et al. 2014). The studies have demonstrated that WWA can:

- monitor consumption of psychoactive substances on campuses servicing thousands of students;
- anonymise locations to avoid ethical risks; and
- incorporate different parts of university complexes, including dormitories to help control for the periods when students are not on campus (e.g. Heuett et al. 2015).

Interestingly Moore et al.'s (2014) study combined survey methods with WWA to link increases in non-therapeutic stimulant consumption with self-reported levels of stress experienced by students. This interdisciplinary approach is a promising example of how WWA sampling can be synchronised with surveys and other methods to provide a range of subjective and objective data.

In terms of ethical issues, WWA studies at universities present fewer risks than prisons, particularly because adult students do not constitute a vulnerable group. Ettore Zuccato and colleagues (2017) conducted a thorough study of eight Italian secondary schools, servicing students aged between 15 and 19 years. Because of heightened ethical risks associated with studying minors (Hall et al. 2012), the team secured the consent of the mayors of the cities involved, water agencies, school authorities and school principals (Zuccato et al. 2017, 286). To control for (a) periods when students were not at school (evenings and weekends) and (b) the contribution of teachers to the wastewater samples, the researchers increased the sample size (approximately 6,126 students) and repeated sampling over a four-year period.

The results demonstrated the capacity of WWA to monitor consumption of multiple drugs, including NPS (e.g. mephedrone). However the findings indicated that illicit drug consumption was largely restricted to cannabis. Levels varied between schools, but consumption was comparable to or lower than in the general population.

The careful project management required to study secondary school students may be daunting for other WWA researchers, particularly if substance use in this age group is limited to cannabis, as indicated by survey data (Hall et al. 2012). Nonetheless, we think it is feasible that an ethically sound WWA project could yield innovative information about the effectiveness of programs to reduce young people's use of tobacco, alcohol or illicit drugs. The research designs could involve the types of empirical approaches discussed above (5.4.3), such as natural experiments, the use of comparator schools as proxy controls, and withdrawal and reversal designs.

5.5.2 Workplaces

Substance use in the workplace is a concern in a variety of settings because of risks associated with integrity (e.g. among police officers), safety and productivity (Allen, Prichard and Griggs 2013). The potential for significant social, economic and environmental harm are also evident in disasters such as the *Exxon Valdez* oil spill in 1989, which may have been partly caused by alcohol intoxication (Liszka 2010).

Strategies that workplaces can use to mitigate these risks include random drug testing, rehabilitation programs and peer-based initiatives (Miller, Zaloshnja and Spicer 2007; Spell and Blum 2005). There is a broad body of literature on the ethical and legal implications of workplace alcohol and drug strategies that interrelate with privacy (e.g. Forrest 1997), discrimination and employment law (e.g. Rothstein 1987), among other things.

Without limiting the possible applications of WWA to workplaces, we suggest that the method could be particularly beneficial in circumstances where:

(a) employees reside on-site and sewerage infrastructure permits sampling of the entire workforce without any contribution from other sources; and
(b) significant harms could arise from accidents caused by intoxication.

Residential mining camps are one example. Some companies have banned alcohol entirely from their mine camps and do not serve alcohol in mess halls (Orr 2014). Theoretically WWA could determine the degree to which such methods have reduced consumption of alcohol. They could in addition detect the use of diverted pharmaceuticals, illicit drugs and substances that cannot be detected by standard urinalysis tests, such as

NPS. We think that WWA may also be used to evaluate the effectiveness of demand reduction strategies, such as peer-based initiatives.

5.6 Conclusion

It is too early to predict whether WWA will play a significant role in improving policy and practices within prison systems. As noted in the Introduction, corrective service agencies may have no interest in changing policies, much less in facilitating novel WWA research. In other settings, internationally recognised ethical principles governing human research may lead researchers to conclude that WWA cannot proceed, irrespective of the interest of prison authorities. Insurmountable technical problems associated with sampling may be encountered.

Nonetheless, unfortunately many prisons exist globally. They house some of the most disadvantaged members of society, including those at risk of death or BBV from drug use. And it seems inevitable that drugs will enter most prisons – decreasing prospects of inmates' successful reintegration post-release and increasing various risks for inmates, staff and the wider community.

A handful of prison WWA studies have shown that application of the method is feasible. This chapter has suggested that under the right technical conditions WWA could be used in sophisticated empirical designs to produce data of a quality that cannot be obtained using other methods – to evaluate drug-related harm and the effects of demand and supply reduction strategies. Similar sorts of empirical designs could be applied to educational institutions and workplaces. However, arguably the need for WWA methods is less compelling in schools and workplaces than in the prison setting and ethical risks remain significant, particularly for studies of school-aged children.

Notes

1. The census uses the term 'confinement institutions' which incorporates prisons, prison farms, correctional centres and work camps.
2. This is not to suggest that the lack of consent would be irrelevant. Arguably it would be an additional relevant consideration, particularly where doubts existed about capacity to satisfy the principles of beneficence and non-maleficence.
3. Indeed, we have found Australian HRECs differed in their opinions as to whether general population studies require oversight.

References

Akers, R. L., N. S. Hayner, and W. Gruninger. 1974. Homosexual and drug behavior in prison: A test of the functional and importation models of the inmate system. *Social Problems* 21:410–22.

Albizu-García, C. E., A. Hernández-Viver, J. Feal, and J. F. Rodríguez-Orengo. 2009. Characteristics of inmates witnessing overdose events in prison: Implications for prevention in the correctional setting. *Harm Reduction Journal* 6, no. 15 (July): 1–8. https://harmreductionjournal.biomedcentral.com/articles/10.1186/1477-7517-6-15.

Allen, J. G., J. Prichard, and L. Griggs. 2013. A workplace drug testing act for Australia. *University of Queensland Law Journal* 32:219–35.

Australian Bureau of Statistics. 2018. 4517.0 – Prisoners in Australia, 2018: Aboriginal and Torres Strait Islander prisoner characteristics. www.abs.gov.au/ausstats/abs@.nsf/Lookup/by%20Subject/4517.0~2018~Main%20Features~Aboriginal%20and%20Torres%20Strait%20Islander%20prisoner%20characteristics%20~13 (accessed 27 March 2020).

Australian Bureau of Statistics. 2019. 4517.0 – Prisoners in Australia, 2019. www.abs.gov.au/ausstats/abs@.nsf/mf/4517.0 (accessed 20 November 2019).

Australian Institute of Health and Welfare. 2019. *The health of Australia's prisoners 2018*. Cat. no. PHE 246. Canberra: Australian Institute of Health and Welfare.

Australian National Council on Drugs. 2008. *National Corrections Drug Strategy 2006–2009*. Canberra: National Drugs Strategy.

BBC News. 2018. Gang who flew drones carrying drugs into prisons jailed. www.bbc.com/news/uk-england-45980560 (accessed 18 November 2018).

Bellair, P. E., and J. E. Sutton. 2018. The reliability of drug use indicators collected from a prisoner sample using the life events calendar method. *Addiction Research & Theory* 26:95–102.

Bijlsma, L., E. Emke, F. Hernández, and P. de Voogt. 2012. Investigation of drugs of abuse and relevant metabolites in Dutch sewage water by liquid chromatography coupled to high resolution mass spectrometry. *Chemosphere* 89:1399–406.

Black E., K. Dolan, and A. Wodak. 2004. *Supply, demand and harm reduction strategies in Australian prisons: Implementation, cost and evaluation*. ANCD research paper 9. Canberra: Australian National Council on Drugs.

Boys, A., M. Farrell, P. Bebbington et al. 2002. Drug use and initiation in prison: Results from a national prison survey in England and Wales. *Addiction* 97:1551–60.

Bresnan, A. 2015. Needle and syringe program opportunity lost for the ACT. www.smh.com.au/opinion/needle-and-syringe-program-opportunity-lost-for-the-act-20150415-1mljfy.html.

Brewer, A. J., C. J. Banta-Green, C. Ort, A. E. Robel, and J. Field. 2016. Wastewater testing compared with random urinalyses for the surveillance of illicit drug use in prisons. *Drug and Alcohol Review* 35:133–37.

Bronson, J., J. Stroop, S. Zimmer, and M. Berzofsky. 2017. *Drug use, dependence, and abuse among state prisoners and jail inmates, 2007–2009*. NCJ250546. Bureau of Justice Statistics. www.bjs.gov/content/pub/pdf/dudaspji0709.pdf.

Burgard, D., R. Fuller, B. Becker, R. Ferrell, and M. Dinglasan-Panlilio. 2013. Potential trends in attention deficit hyperactivity disorder (ADHD) drug use on a college campus: Wastewater analysis of amphetamine and ritalinic acid. *Science of the Total Environment* 450–451:242–49.

Carson, E. A. 2018. *Prisoners in 2016*. NCJ 251149. Bureau of Justice Statistics. www.bjs.gov/content/pub/pdf/p16.pdf.

Cope, N. 2000. Drug use in prison: The experience of young offenders. *Drugs: Education, Prevention and Policy* 7:355–66.

Crewe, B. 2005. Prisoner society in the era of hard drugs. *Punishment & Society* 7:457–81.

Crime and Corruption Commission (CCC). 2018. *Taskforce Flaxton: An examination of corruption risks and corruption in Queensland prisons*. Brisbane: Crime and Corruption Commission.

Csete, J., A. Kamarulzaman, M. Kazatchkine et al. 2016. Public health and international drug policy. *The Lancet* 387:1427–80.

Dean, J. 2005. The future of mandatory drug testing in Scottish prisons: A review of policy. *International Journal of Prisoner Health* 1:163–70.

Dolan, K., E. M. Khoei, C. Brentari, and A. Stevens. 2007. *Prisons and drugs: A global review of incarceration, drug use and drug services*. https://beckleyfoundation.org/wp-content/uploads/2016/04/BF_Report_12.pdf.

Dolan, K., and A. Rodas. 2014. Detection of drugs in Australian prisons: Supply reduction strategies. *International Journal of Prisoner Health* 10:111–17.

European Monitoring Centre for Drugs and Drug Addiction. 2012. *Prisons and drugs in Europe: The problem and responses*. Luxembourg: Publications Office of the European Union.

European Monitoring Centre for Drugs and Drug Addiction. 2014. *Estimating public expenditure on drug-law offenders in prison in Europe*. EMCDDA Papers. Luxembourg: Publications Office of the European Union.

Fazel, S., P. Bains, and H. Doll. 2006. Substance abuse and dependence in prisoners: A systematic review. *Addiction* 101:181–91.

Forrest, A. R. W. 1997. Ethical aspects of workplace urine screening for drug abuse. *Journal of Medical Ethics* 23:12–17.

Gast, D. L., J. R. Ledford, and K. E. Severini. 2018. Withdrawal and reversal designs. In *Single case research methodology: Applications in special education and behavioral sciences*, ed. J. R. Ledford, and D. L. Gast, 215–38. New York: Routledge.

Gillespie, W. 2005. A multi-level model of drug abuse inside prison. *The Prison Journal* 85:223–46.

Gore, S. M., and A. G. Bird. 1996. Cost implications of random mandatory drugs tests in prisons. *The Lancet* 348:1124–27.

Gushgari, A. J., E. M. Driver, J. C. Steele, and R. U. Halden. 2018. Tracking narcotics consumption at a Southwestern U. S. university campus by wastewater-based epidemiology. *Journal of Hazardous Materials* 359:437–44.

Hall, W., J. Prichard, P. Kirkbride et al. 2012. An analysis of ethical issues in using wastewater analysis to monitor illicit drug use. *Addiction* 107:1767–73.

Hancock, P., and Y. Jewkes. 2011. Architectures of incarceration: The spatial pains of imprisonment. *Punishment & Society* 13:611–29.

Harrison, H. 1997. The validity of self-reported drug use in survey research: An overview and critique of research methods. In *The validity of self-reported drug use: Improving the accuracy of survey estimates*, ed. L. Harrison, and A. Hughes, 17–36. Rockville, MD: National Institute on Drug Abuse.

Her Majesty's Inspectorate of Prisons. 2019. *Her Majesty's Chief Inspector of Prisons for England and Wales: Annual Report 2018–19*. London: Her Majesty's Inspectorate of Prisons.

Heuett, N.V., C.E. Ramirez, A. Fernandez, and P.R. Gardinali. 2015. Analysis of drugs of abuse by online SPE-LC high resolution mass spectrometry: Communal assessment of consumption. *Science of The Total Environment* 511:319–30.

Hughes, R. 2000. Drug injectors and prison mandatory drug testing. *The Howard Journal of Criminal Justice* 39:1–13.

Jacques, O. 2019. Ex-prisoners break parole to smuggle drugs into Queensland prisons, union claims. ABC News. www.abc.net.au/news/2019-07-10/drugs-being-smuggled-into-prisons/11289084 (accessed 25 November 2019).

Kaeble, D. and M. Cowhig. 2018. *Correctional populations in the United States*. NCJ 251211. Bureau of Justice Statistics. www.bjs.gov/content/pub/pdf/cpus16.pdf (accessed 25 March 2019).

Kang, S., S. Deren, J. Andia, H. M. Colón, R. Robles, and D. Oliver-Velez. 2005. HIV transmission behaviours in jail/prison among Puerto Rican drug injectors in New York and Puerto Rico. *AIDS and Behaviour* 9:377–86.

Kendall P. R. W., and M. Pearce. 2000. Drug testing in Canadian jails: To what end? *Canadian Journal of Public Health* 91:26–28.

Kevin, M. 2013. *Drug use in the inmate population – Prevalence, nature and context*. Corrective Services NSW. www.correctiveservices.justice.nsw.gov.au/Documents/Related%20Links/publications-and-policies/cres/research-publications/drug-use-in-the-inmate-population-prevalence-nature-and-context.pdf (accessed 27 March 2020).

Lai, F. Y., P. K. Thai, J. O'Brien et al. 2013. Using quantitative wastewater analysis to measure daily usage of conventional and emerging illicit drugs at an annual music festival. *Drug and Alcohol Review* 32:594–602.

Lancaster, K., T. Rhodes, K. Valentine, and A. Ritter. 2019. A 'promising tool'? A critical review of the social and ethico-political effects of wastewater analysis in the context of illicit drug epidemiology and drug policy. *Current Opinion in Environmental Science & Health* 9:85–90.

Lancaster, K., A. Ritter, and T. Rhodes. 2019. 'A more accurate understanding of drug use': A critical analysis of wastewater analysis technology for drug policy. *International Journal of Drug Policy* 63:47–55.

Larney, S., N. Gisev, M. Farrell, T. Dobbins, L. Burns, A. Gibson, J. Kimber, and L. Degenhardt. 2014. Opioid substitution therapy as a strategy to reduce deaths in prison: Retrospective cohort study. *BMJ Open* 4:1–8.

Levy, M. H., C. Treloar, R. M. McDonald, and N. Booker. 2007. Prisons, hepatitis C and harm minimisation. *The Medical Journal of Australia* 186:647–9.

Link, B. G., and J. C. Phelan. 2001. Conceptualizing stigma. *Annual Review of Sociology* 27:363–85.

Liszka, J. 2010. Lessons from the Exxon Valdez oil spill: A case study in retributive and corrective justice for harm to the environment. *Ethics and the Environment* 15:1–30.

Loxley, W., J. W. Toumbourou, T. Stockwell et al. 2004. *The prevention of substance use, risk and harm in Australia: A review of the evidence*. Canberra: The National Drug Research Centre and the Centre for Adolescent Health.

MacPherson, P. 2004. *Use of random urinalysis to deter drug use in prison: A review of the issues*. Ottawa: Correctional Service of Canada.

Marlowe, D. B. 2011. Evidence-based policies and practices for drug-involved offenders. *The Prison Journal* 91:27–47.

McConnell, A., and P. 't Hart. 2019. Inaction and public policy: Understanding why policymakers 'do nothing'. *Policy Sciences* 52:645–61.

McGregor, K., and T. Makkai. 2003. Self-reported drug use: How prevalent is underreporting?. *Trends and Issues in Criminal Justice* 260:1–6.

Metzner, J. L., and J. Fellner. 2010. Solitary confinement and mental illness in U. S. prisons: A challenge for medical ethics. *Journal of the American Academy of Psychiatry and the Law Online* 38:104–8.

Miller, T. R., E. Zaloshnja, and R. S. Spicer. 2007. Effectiveness and benefit-cost of peer-based workplace substance abuse prevention coupled with random testing. *Accident Analysis & Prevention* 39:565–73.

Ministry of Justice. 2019. *Offender Management Statistics Bulletin, England and Wales.* https://assets.publishing.service.gov.uk/government/uploads/system/uploads/attachment_data/file/842590/OMSQ_2019_Q2.pdf (accessed 20 November 2019).

Moore, D. R., D. A. Burgard, R. G. Larson, and M. Ferm. 2014. Psychostimulant use among college students during periods of high and low stress: an interdisciplinary approach utilizing both self-report and unobtrusive chemical sample data. *Addictive Behaviors* 39:987–93.

Moore, K. E., J. B. Stuewig, and J. P. Tangney. 2016. The effect of stigma on criminal offenders' functioning: A longitudinal mediational model. *Deviant Behavior* 37:196–218.

Mundt, A. P., G. Baranyi, C. Gabrysch, and S. Fazel. 2018. Substance use during imprisonment in low-and middle-income countries. *Epidemiologic Reviews* 40:70–81.

National Health and Medical Research Council, Australian Research Council, and Universities Australia. 2007. *National Statement on Ethical Conduct in Human Research 2007 (Updated 2018).* Canberra: Commonwealth of Australia.

Néfau, T., O. Sannier, C. Hubert, S. Karolak, and Y. Lévi. 2017. *Analysis of drugs in sewage: An approach to assess substance use, applied to a prison setting.* https://en.ofdt.fr/BDD/publications/docs/eisatnx3.pdf.

Nunn, A., N. Zaller, S. Dickman, C. Trimbur, A. Nijhawan, and J. D. Rich. 2009. Methadone and buprenorphine prescribing and referral practices in US prison systems: Results from a nationwide survey. *Drug and Alcohol Dependence* 105:83–88.

Orr, A. 2014. WA's Argyle Diamond Mine wet mess goes dry. *Sydney Morning Herald.* www.smh.com.au/business/companies/was-argyle-diamond-mine-wet-mess-goes-dry-20140704-zswqc.html.

Ort, C., M. G. Lawrence, J. Rieckermann, and A. Joss. 2010. Sampling for pharmaceuticals and personal care products (PPCPs) and illicit drugs in wastewater systems: Are your conclusions valid? A critical review. *Environmental Science and Technology* 44:6024–35.

Panawennage, D., S. Castiglioni, E. Zuccato, E. Davoli, and M. P. Chiarelli. 2011. Measurement of illicit drug consumption in small populations: Prognosis for noninvasive drug testing of student populations. In *Illicit drugs in the environment: Occurrence, analysis, and fate using mass spectrometry*, ed. S. Castiglioni, E. Zuccato, and R. Fanelli, 321–331. New Jersey: John Wiley and Sons.

Paul, N. W., A. Caplan, M. E. Shapiro, C. Els, K. C. Allison, and H. Li. 2017. Human rights violations in organ procurement practice in China. *BMC Medical Ethics* 18, no. 1 (February): 1–9.

Penfold, C., P. J. Turnbull and R. Webster. 2005. *Tackling prison drug markets: An exploratory study.* London: Great Britain Home Office Research Development and Statistics Directorate.

Plourde, C., S. Brochu, A. Gendron, and N. Brunelle. 2012. Pathways of substance use among female and male inmates in Canadian federal settings. *The Prison Journal* 92:506–24.

Poehlmann, J., D. Dallaire, A. B. Loper, and L. D. Shear. 2010. Children's contact with their incarcerated parents: Research findings and recommendations. *American Psychologist* 65:575–98.

Postigo, C., M. López de Alda, and D. Barceló. 2011. Evaluation of drugs of abuse use and trends in a prison through wastewater analysis. *Environment International* 37:49–55.

Prendergast, M. L., M. Campos, D. Farabee, W. K. Evans, and J. Martinez. 2004. Reducing substance use in prison: The California drug reduction strategy project. *The Prison Journal* 84:265–80.

Prichard, J., W. Hall, E. Zuccato et al. 2016. *Ethical research guidelines for sewage epidemiology*. www.emcdda.europa.eu/system/files/attachments/10405/WBEethical-guidelines-v1.0- 03.2016.pdf (accessed 2 December 2017).

Prichard, J., C. Ort, R. Bruno et al. 2010. Developing a method for site-specific wastewater analysis: Implications for prisons and other agencies with an interest inillicit drug use. *Journal of Law, Information and Science* 20:15–27.

Prichard, J., and J. Payne. 2005. *Alcohol, drugs and crime: A study of juveniles in detention*. Research and Public Policy Series no 67. Canberra: Australian Institute of Criminology.

Ralphs, R., L. Williams, R. Askew, and A. Norton. 2017. Adding spice to the porridge: The development of a synthetic cannabinoid market in an English prison. *International Journal of Drug Policy* 40:57–69.

Richters, J., T. Butler, L. Yap et al. 2008. *Sexual health and behaviour of New South Wales prisoners*. Sydney: School of Public Health and Community Medicine, University of New South Wales.

Rodas, A., A. Bode, and K. Dolan. 2012. *Supply, demand and harm reduction strategies in Australian prisons: An update*. Canberra: Australian National Council on Drugs.

Rothstein, M. A. 1987. Drug testing in the workplace: The challenge to employment relations and employment law. *Chicago-Kent Law Review* 63: 683–743.

Royuela, L., L. Montanari, M. Rosa, and J. Vicente. 2014. *Drug use in prison: Assessment report – Reviewing tools for monitoring illicit drug use in prison populations in Europe*. European Monitoring Centre for Drugs and Drug Addiction. www.emcdda.europa.eu/system/files/publications/784/Drug_use_in_prison_assessment_report_462763.pdf.

Schofield, P., T. Butler, S. Hollis, and C. D'Este. 2011. Are prisoners reliable survey respondents? A validation of self-reported traumatic brain injury (TBI) against hospital medical records. *Brain Injury* 25:74–82.

Shearer, J., B. White, S. Gilmour, A. D. Wodak, and K. A. Dolan. 2006. Hair analysis underestimates heroin use in prisoners. *Drug and Alcohol Review* 25:425–31.

Souryal, S. S. 2009. Deterring corruption by prison personnel: A principle-based perspective. *The Prison Journal* 89:21–45.

Spell, C. S., and T. C. Blum. 2005. Adoption of workplace substance abuse prevention programs: Strategic choice and institutional perspectives. *Academy of Management Journal* 48:1125–42.

Steels, B., and D. Goulding. 2009. *Predator or prey? An exploration of the impact and incidence of sexual assault in West Australian Prisons*. Perth: Centre for Social and Community Research, Murdoch University.

Stephan, J. J. 2008. *Census of state and federal correctional facilities, 2005*. NCJ 22218. Bureau of Justice Statistics. www.bjs.gov/content/pub/pdf/csfcf05.pdf

Stöver, H., and J. Nelles. 2003. Ten years of experience with needle and syringe exchange programmes in European prisons. *International Journal of Drug Policy* 14:437–44.

Sykes, G. M. 1958 (2007). *The society of captives: A study of a maximum security prison*. New Jersey: Princeton University Press (Reprint).

United Nations Office on Drugs and Crime. 2015. *Handbook on dynamic security and prison intelligence*. New York: United Nations.

United Nations Office on Drugs and Crime. 2017. *Handbook on anti-corruption measures in prisons*. Vienna: United Nations.

United Nations Office on Drugs and Crime. 2019. *World drug report 2019 (Booklet 2 – Global overview of drug demand and supply)*. New York: United Nations.

van Dyken, E., F. Y. Lai, P. K. Thai et al. 2016. Challenges and opportunities in using wastewater analysis to measure drug use in a small prison facility. *Drug and Alcohol Review* 35:138–47.

van Dyken, E., P. Thai, F. Y. Lai et al. 2014. Monitoring substance use in prisons: Assessing the potential value of wastewater analysis. *Science and Justice* 54:338–45.

Vandam, L. 2009. Patterns of drug use before, during and after detention: A review of epidemiological literature. In *Contemporary issues in the empirical study of crime*, ed. M. Cools, S. De Kimpe, B. De Ruyver et al., 227–55. Antwerp, Belgium: Apeldoorn; The Netherlands: Maklu.

Walmsley, R. 2018. *World prison population list* (12th ed). www.prisonstudies.org/sites/default/files/resources/downloads/wppl_12.pdf (accessed 25 March 2019).

Winfree, L. T., G. Newbold, and S. H. Tubb III. 2002. Prisoner perspectives on inmate culture in New Mexico and New Zealand: A descriptive case study. *The Prison Journal* 82:213–33.

Winnick, T. A., and M. Bodkin. 2008. Anticipated stigma and stigma management among those to be labeled 'Ex-Con'. *Deviant Behavior* 29:295–333.

Wood, E., J. Montaner, and T. Kerr. 2005. HIV risks in incarcerated injection-drug users. *The Lancet* 366:1834–35.

Woodall, J. 2012. Social and environmental factors influencing in-prison drug use. *Health Education* 112:31–46.

Zuccato, E., E. Gracia-Lor, N. I. Rousis et al. 2017. Illicit drug consumption in school populations measured by wastewater analysis. *Drug and Alcohol Dependence* 178:285–90.

chapter six

Future Directions

INTRODUCTION

Preceding chapters have provided what we think are well-grounded recommendations for how current WWA technology could be applied to augment existing research methods on drug use, such as surveys. We have focussed mainly on illicit drugs, but we have also discussed the use of WWA to monitor population use of tobacco, alcohol, new psychoactive substances and diverted pharmaceuticals. It is worth summarising some of the key recommendations from each chapter and their intended audiences. This is the focus of 6.1. The following sections discuss broader issues relating to the future of WWA.

6.1 Overview of Recommended Future Applications of WWA

We hope that Chapter 3 will interest global agencies, including the World Health Organization (WHO) and the United Nations Office on Drugs and Crime (UNODC), in the value of WWA in working to reduce the global burden of disease attributable to psychoactive substance use. We highlighted that the metrics used to inform international policy – surveys and event data – are weakest in lower- and middle-income countries (LMIC) where the need for better metrics is greatest because of the burdens LMIC populations endure from consuming tobacco, alcohol and other drugs. The resources and skills required to gather and analyse survey data to estimate consumption in LMIC are scant.

WWA can potentially analyse wastewater samples from LMICs by targeting sewerage treatment plants (WWTPs) that have adequate environments for sampling. We have not certified that such WWTPs exist, but it seems likely given publicly available information, for example, Kenya's Dandora WWTP, Jordan's As-Samra WWTP and India's extensive network of over 500 WWTPs. Local water authorities can follow sampling procedures, which are critical, but uncomplicated and comparatively quick. Samples can be sent to laboratories for analysis to provide descriptive data about the consumption of psychoactive substances relevant to the country of origin and its region (e.g. sub-Saharan Africa). With a low investment of time and resources, policy makers could have access to annual time series about regions where no data are now available.

Chapter 4 was directed at geographically large countries, such as America and Australia, where information about rural populations is of a lower quality than that in urban settings, which are easier to study using surveys. Annually collected WWA data from rural sites will improve the evidence base for policy in rural communities. One-off WWA studies in particular rural towns may be useful to address public concerns – well founded or otherwise – about spikes in consumption of specific drugs. We also recommended that scholars interested in rural populations consider how WWA may be used to conduct interdisciplinary evaluative studies, such as natural experiments and intervention studies that could evaluate the effects of policies on drug markets (Babor et al. 2010). Ironically, the isolation of some rural towns – which often inhibits research – would make them superior to urban settings for WWA-based policy evaluations.

The ethical complexities associated with WWA research increase as its empirical focus narrows from very large populations, to small rural towns, to specific facilities like prisons and schools. For this reason, Chapter 5 argued that the future of WWA prison research is uncertain. Our primary recommendation was that WWA researchers exercise due diligence, ideally with the guidance of human research ethics committees (HRECs), to assess the ethical risks associated with prison research. These issues may be more complicated in doing research on prisons based in a different country to the researchers. Where ethical risks are sufficiently low and risk mitigation strategies are in place, WWA may be able to improve the lot of prisoners (and others) through monitoring drug markets and evaluating strategies designed to reduce drug-related harms and to reduce supply and demand.

Based on her experience advocating for inmates' rights in North America, one colleague warned us that WWA can only be used in ways that will reduce inmates' quality of life. We appreciate that there are contexts where this is likely to be true. However, we are not convinced that this is universally true. And so without offering a definitive recommendation in favour of prison studies, we think it is feasible that there are prison settings in which the potential benefits of WWA outweigh the risks involved in the research.

6.2 Extending Ethical Considerations

Several broader points should be emphasised about the future of WWA. The WWA field has built a good reputation for several reasons: its demonstrable scientific rigour, its efforts to address pressing real-world problems, and the fact that researchers have exercised caution on ethical issues and sought advice. Insofar as media reports reflect community views, it seems that the public broadly accept that the risks to individuals arising from WWA are negligible and that this type of research is useful.

Currently other disciplines are accepting of WWA, or at least are not actively hostile towards it. However, reputations of research fields can be damaged. For WWA potentially this could occur if the poor conceptualisation of ethical risks leads to negative outcomes. Foreseeable harms to different groups of people were underscored in Chapters 3 to 5 (e.g. stigmatisation of vulnerable groups and the misinterpretation of WWA data leading to poor policy outcomes).

Depending on the nature of the harms caused by illicit drugs, we can envisage negative media coverage of WWA research that elicits the disapproval of other disciplines involved in drug research, undermines public distrust or increases anger, and reduces the interest in and funding of WWA research from policy makers. Negative perceptions of WWA could tap into fears about the growth in the surveillance powers of the state arising from technological advances in other areas (e.g. genomics, social media, 'big data', facial recognition and so on). Unlike survey data, with which the public is more familiar, WWA has the potential to become, or be perceived as, an Orwellian tool – a means by which the state can spy on its citizens and promote 'tough on drugs' law enforcement approaches.

Some of these concerns were raised by Lancaster, Ritter and Rhodes (2019) in their critique of the ways in which they claim that WWA data are constructed, particularly by agencies such as the Australian Criminal Intelligence Commission and the European Monitoring Centre for Drugs and Drug Addiction (EMCDDA). Among other things, these authors argued that:

- the language used within WWA literature to describe problems associated with survey data stigmatise people who use illicit drugs as inherently "criminal, untrustworthy, deviant and in need of surveillance"; and
- WWA literature privileges "objective scientific data over other self-report methods in the drugs field" (Lancaster, Ritter and Rhodes 2019, 51, 53).

We find the first claim puzzling. The content and quality of illicit drugs is uncertain because of adulteration and substitution so people who use them may not be aware what substances they are consuming. Indeed, this fact provides the rationale for drug checking which several of these authors strongly support. WWA does emphasise the objectivity of data on the nature of illicit drugs that are used but it does not 'privilege' these data over self-reported patterns of drug use. Indeed, these are essential in understanding the trends in population drug use that are revealed by WWA.

We agree that WWA researchers should pay close attention to how findings are interpreted, described and acted upon by law enforcement and other agencies. They should accordingly be ready to publicly rectify misrepresentations of WWA data on drug consumption.

However, the cogency of Lancaster, Ritter and Rhodes' (2019) arguments was undermined in a number of ways. First, Lancaster, Ritter and Rhodes (2019) did not consider the WWA ethics literature on the stigmatisation of vulnerable groups (e.g. Hall et al. 2012). This was acknowledged to some extent in a subsequent publication (Lancaster, Rhodes, Valentine et al. 2019).

The authors also discounted the fact that the WWA literature has consistently presented WWA as a way of augmenting rather than replacing other methods such as surveys. Some publications have used WWA data to critically examine political and media narratives about methamphetamine consumption (e.g. Prichard et al. 2018). Finally, many of the authors' criticisms of WWA also apply to the use of other scientific data that have been used in long-standing drug monitoring systems, including three such data collections in Australia (see Prichard et al. 2017, 839–841).

Turning to other issues, an area that will need careful future examination by the WWA field is the use of general population studies in jurisdictions that have particularly severe, if not draconian, drug policies. Many jurisdictions in high income countries could be criticised for drug policies that have failed to strike a reasonable balance between reducing harm, demand and supply (e.g. Csete et al. 2016; Rodas, Bode and Dolan 2012). However, the policies of some jurisdictions raise weighty ethical issues that are qualitatively different to those in most Western countries, such as those studied by the EMCDDA. To point to two examples, we are not confident that the principles of beneficence and non-maleficence (see 2.3) can be satisfied in jurisdictions where capital punishment is used for drug trafficking (see Girelli 2019), or where coercive measures are used against people with substance use disorders, such as administrative detention or labour camps (e.g. Yuan 2019, 484). The WWA ethical guidelines (Prichard et al. 2016) will need to be updated to incorporate analysis of these issues as the use of WWA expands beyond high-income countries.

While this book has concentrated on WWA teams as the drivers of research, it may be the case that third parties approach researchers for advice on how to do studies, or ask them to undertake fee-for-service work. The advice or work may not constitute 'research' in some definitions (e.g. see NHMRC 2007 (updated 2018), 7–8) – noting that human research ethics work in different ways around the world. However, it may be the case that the contribution of the WWA team – paid or unpaid – *does* constitute research to which human research ethical principles apply *even though the team does not intend to publish the data* (e.g. see NMHRC 2007 (updated 2018), 8). If this is the case, then the ethical risks of consultancies may be greater precisely because the WWA team have less ability to influence the study, including the capacity to abandon research altogether. By way of example, a private consultancy may seek to profit from corrective service authorities by conducting WWA with the collaboration of WWA researchers to oversee sampling and to conduct laboratory analyses. In such circumstances HREC oversight would be wise (if not mandatory).

The scientific quality of contract research is another issue of concern. Currently relatively few researchers have expertise to (a) conduct sampling, especially from buildings, and (b) analyse samples in properly equipped laboratories. Therefore government and non-government agencies should be wary of contracting the services of consultants who purport to be able to conduct WWA with unsophisticated methods. They may not understand issues such as how to do back-calculations (see 2.2.3), or the problems associated with 'grab samples' (see 2.2.2). Alternatively they may rely on commercial laboratories that use analytical equipment that lacks the sensitivity required to produce reliable estimates. The data produced by such methods may contain significant errors and result in dramatic over- or underestimates of consumption.

6.3 *Developments in Infrastructure and Technology*

The fact that multiple research disciplines study psychoactive substance use is understandable given the size of the global health problems it causes and the capacity of illicit drug markets to threaten security, sustainable development, good governance and economic stability (UNODC 2016b; see 1.1). This book has analysed opportunities for WWA to contribute to research on some of these issues at international, national and local levels. However, parallel to many of those opportunities are additional applications of WWA that are unrelated to the use of psychoactive substances. These were briefly touched upon earlier. They include the capability for WWA to study:

- indicators of human health relating to diet, cancer, physical trauma and exposure to pollutants (see 2.2.1); and
- pollution in the natural environment (see 2.2.2).

Because of the techniques that underpin WWA, the *same* samples taken to study consumption of psychoactive substances can also be analysed to examine very many other health and environmental issues. The best way for non-scientists to find out what is feasible is to communicate with WWA scientists. Laboratories need to develop procedures for identifying particular chemicals and interpreting their findings. The capabilities of different laboratories will vary. Laboratories are also likely to be aware of new advances that occur in the WWA field that could generate new research possibilities that are pertinent to policy and research. This includes the types of chemicals that can be examined, as well as developments in completely different processes, such as passive sampling (see EMCDDA 2016).

However, the potential implications of the additional layers of data that WWA could yield are interesting to consider. By way of example, in theory samples taken from multiple WWTPs in India could be analysed to monitor:

- aspects of community diet;
- rates of certain diseases;
- population-level consumption of tobacco, alcohol and illicit drugs; and
- the degree to which WWTPs are abiding by health and environmental regulations by adequately treating wastewater.

The latter could be achieved by comparing the contents of *influent* with *effluent* – mirroring processes already in place in Australia. Similar concepts may be worth pursuing in studies of community and environmental wellbeing in rural areas in Australia, America and other countries, along the lines proposed in Chapter 4.

Earlier we noted that sewerage treatment infrastructure can be conceptualised as the data collection apparatus of WWA (2.2.2). As Daughton (2018) warned, changes to the engineering systems used by water authorities could present empirical challenges to WWA. It is feasible that over time the number of population areas where WWA can be conducted might contract, particularly as the pressure to conserve water increases. Of course, it is also true that changes to infrastructure may benefit WWA research by increasing the numbers of conventional WWTPs to sample, or by designing sewers that facilitate optimal sampling conditions, as outlined in Chapter 5. On a separate note, we wonder whether fruitful new areas of WWA could explore the feasibility of sampling from some of the non-conventional sewerage systems that are found in LMICs, including types of decentralised wastewater treatment systems (see e.g. UN and AIT 2015).

6.4 Conclusion

Clearly there are many potential new beneficial applications of WWA and means by which it may be able to assist to provide a fuller picture for researchers and policy makers. Predicting which developments will succeed is a difficult task. And it is very likely that major contributions will be achieved that surprise researchers including ourselves.

But the process of writing this book has affirmed the ideas we set out in Chapter 1 regarding the benefits of multi-disciplinary collaboration. For

WWA researchers we commend collaboration with multiple academic disciplines, policy makers, practitioners and HREC to: (a) assist in generating research of maximum civic value; (b) better interpret study findings; (c) conduct mixed method approaches (e.g. incorporating quantitative and qualitative social science methods); (d) increase the chances of uncovering errors in research; and (e) wisely adhere to ethical principles governing human research. We hope this book will encourage others to consider collaborating with WWA teams in the knowledge that their own expertise is likely to benefit the field.

References

Babor, T. F., J. P. Caulkins, G. Edwards et al. 2010. *Drug policy and the public good*. Oxford: Oxford Univ. Press.

Csete, J., A. Kamarulzaman, M. Kazatchkine et al. 2016. Public health and international drug policy. *The Lancet* 387:1427–80.

Daughton, C. G. 2018. Monitoring wastewater for assessing community health: Sewage Chemical-Information Mining (SCIM). *Science of The Total Environment* 619–620:748–64.

European Monitoring Centre for Drugs and Drug Addiction. 2016. *Assessing illicit drugs in wastewater: Advances in wastewater-based drug epidemiology*. Insights 22. Luxembourg: Publications Office of the European Union.

Girelli, G. 2019. *The death penalty for drug offences: Global overview 2018*. London: Harm Reduction International.

Hall, W., J. Prichard, P. Kirkbride et al. 2012. An analysis of ethical issues in using wastewater analysis to monitor illicit drug use. *Addiction* 107:1767–73.

Lancaster, K., T. Rhodes, K. Valentine, and A. Ritter. 2019. A 'promising tool'? A critical review of the social and ethico-political effects of wastewater analysis in the context of illicit drug epidemiology and drug policy. *Current Opinion in Environmental Science & Health* 9:85–90.

Lancaster, K., A. Ritter, and T. Rhodes. 2019. 'A more accurate understanding of drug use': A critical analysis of wastewater analysis technology for drug policy. *International Journal of Drug Policy* 63:47–55.

National Health and Medical Research Council, Australian Research Council, and Universities Australia. 2007. *National Statement on Ethical Conduct in Human Research 2007 (updated 2018)*. Canberra: Commonwealth of Australia.

Prichard, J., W. Hall, E. Zuccato et al. 2016. *Ethical research guidelines for sewage epidemiology*. www.emcdda.europa.eu/system/files/attachments/10405/WBE-ethical-guidelines-v1.0–03.2016%20.pdf (accessed 2 December 2017).

Prichard, J., F. Y. Lai, J. O'Brien et al. 2018. 'Ice rushes', data shadows and methylamphetamine use in rural towns: Wastewater analysis. *Current Issues in Criminal Justice* 29:195–208.

Prichard, J., F. Y. Lai, E. van Dyken et al. 2017. Wastewater analysis for estimating substance use: Implications for law, policy and research. *Journal of Law and Medicine* 24:837–49.

Rodas, A., A. Bode, and K. Dolan. 2012. *Supply, demand and harm reduction strategies in Australian prisons: An update*. Canberra: Australian National Council on Drugs.

United Nations Economic and Social Commission for Asia and the Pacific, United Nations Human Settlements Programme and Asian Institute of Technology. 2015. *Policy guidance manual on wastewater management with a special emphasis on decentralized wastewater treatment systems*. Bangkok: United Nations.

United Nations Office on Drugs and Crime. 2016b. *UNODC annual report: Covering activities during 2016*. www.unodc.org/documents/AnnualReport2016/2016_UNODC_Annual_Report.pdf.

Yuan, X. 2019. Controlling illicit drug users in China: From incarceration to community? *Australian & New Zealand Journal of Criminology* 52:483–98.

Index

Note: Page numbers in *italics* indicate figures and in **bold** indicate tables on the corresponding pages.

Lightning Source UK Ltd.
Milton Keynes UK
UKHW020402190522
403164UK00001B/8